Models and Approaches to STEM Professional Development

Models AND Approaches TO STEM Professional Development

Brenda S. Wojnowski and
Celestine H. Pea, Editors

NSTA press
National Science Teachers Association
Arlington, Virginia

Claire Reinburg, Director
Wendy Rubin, Managing Editor
Andrew Cooke, Senior Editor
Amanda O'Brien, Associate Editor
Amy America, Book Acquisitions Coordinator

Art and Design
Will Thomas Jr., Director

Printing and Production
Catherine Lorrain, Director

National Science Teachers Association
David L. Evans, Executive Director
David Beacom, Publisher

1840 Wilson Blvd., Arlington, VA 22201
www.nsta.org/store
For customer service inquiries, please call 800-277-5300.

17 16 15 14 4 3 2 1

Library of Congress Cataloging-in-Publication Data
Models and approaches to STEM professional development / Brenda S. Wojnowski and Celestine H. Pea, editors.
pages cm.
Includes bibliographical references and index.
ISBN 978-1-936137-35-0 -- ISBN 978-1-938946-68-4 (ebook) 1. Science teachers--Training of--United States. 2. Engineering teachers--Training of--United States. 3. Mathematics teachers--Training of--United States. I. Wojnowski, Brenda S., 1948-, editor of compilation. II. Pea, Celestine H., 1946-, editor of compilation.
Q183.3.A1M6659 2013
507.1'173--dc23

2013037231

Cataloging-in-Publication Data for the e-book are also available from the Library of Congress.

CONTENTS

FOREWORD

Patricia M. Shane

This volume arrives at a most propitious time for those involved in science education in the United States. As a nation, we are entering a time of significant transition as we prepare to digest, assimilate, and enact the changes inherent in achieving the goals of the *Next Generation Science Standards* (*NGSS*). These changes allow for a focus on the core ideas in science and engineering as well as their practices and the crosscutting concepts that are common to both dimensions. Integral to the process of change is the need to be removed from the comfort zone of our current practices. Thus, no matter how great the recognition of the need for change, the process remains arduous and stressful—even for the most passionate proponents.

Models and Approaches to STEM Professional Development provides direction to managing the changes entailed in adoption of the new standards. It takes a meaningful look at the history of professional development in science education, discusses challenges of the new standards and related research on learning, highlights critical aspects of successful programs, and provides forward-facing insights into the needed professional development surrounding the *NGSS*.

The case for the importance of science, technology, engineering, and mathematics (STEM) reforms and their relevance to professional development is clearly delineated by the authors. Considerable attention is given to creating new ways of listening to and monitoring students' scientific reasoning and thinking as well as the importance of professional development designed to enact science reforms. Concomitantly, careful blending of what is new, especially *A Framework for K–12 Science Education* and the *NGSS*, with the successes of existing science professional development programs are strengths of this volume. As George Santayana so eloquently said, "Those who cannot remember the past are condemned to repeat it." Because the advent of new standards doesn't mean ignoring successes of the past, wise implementers will embrace those programs that have been successful and build upon them as they embark on new endeavors.

Because it emphasizes the strengths of existing models, this book does an excellent job of sharing the advantages of nine successful science professional development programs across the country. Some are local programs while others are statewide or regional, but they have elements in common such as grassroots efforts, involvement of the players in developing a program, in-depth professional development over time, and formative evaluation to guide ongoing program revision. Further, insights into the sustainability of the programs are detailed. These are all programmatic elements that need to be considered as we embark on the next stage of science education reform in the United States.

FOREWORD

Leaders from across the country have come together in this volume to share their cumulative wisdom about lessons learned. The book demonstrates how new wheels do not have to be invented to enact the *NGSS* and clearly lays out considerations and methodologies for building on current science education wheels while incorporating new research about how students learn. These themes are deftly developed and articulate the appropriate pathways to achieving the goals of the new science standards. The lessons learned from successful programs are provided along with specific examples of what made them thrive. In addition, considerable attention is given to developing new ways of listening to and monitoring students' scientific reasoning and thinking. In sum, this volume combines the best of what we have learned since the advent of science reform in order to prepare us for the transition to the recently released *NGSS*.

ABOUT THE EDITORS

Brenda S. Wojnowski

Brenda Shumate Wojnowski, EdD, is president of a Dallas-based education consulting firm geared toward nonprofit and university clients. She is a past president of the National Science Education Leadership Association (NSELA) and a past chair of the National Science Teachers Association (NSTA) Alliance of Affiliates. Dr. Wojnowski edits the NSELA journal, *Science Educator,* and chaired the 2010 NSTA STEM (science, technology, engineering, and mathematics) task force. She has been engaged in university- and foundation-based programs for over 25 years, with prior experience in public schools. During her career, she has served as senior program officer for a nonprofit foundation and president of a museum-based nonprofit. Dr. Wojnowski has held a variety of university positions, including teaching graduate-level courses in educational leadership and researching and supporting STEM areas. An award-winning K–12 teacher, she has taught at the middle and secondary levels and has served as a high school curriculum administrator. She holds a doctorate in curriculum and teaching with postdoctoral work in educational administration, a master of arts in middle grades education, an undergraduate degree in biology with a minor in secondary education, and teaching and supervision licensures in eight areas. She has presented numerous workshops and invited talks as well as having served in a senior level capacity on many grants and contracts from public agencies and private foundations. Dr. Wojnowski has numerous publications to her credit. Her research interests are in STEM areas, school reform, and the mentoring of beginning teachers.

Celestine H. Pea

Celestine (Celeste) H. Pea, PhD, is a program director in the Division of Research on Learning (DRL), Education and Human Resource Directorate, National Science Foundation. In DRL, Dr. Pea works primarily with the Research on Education and Learning program, for which she manages a portfolio of awards that conduct interdisciplinary research about STEM in current and emerging contexts. She also works with the Innovative Technology Experiences for Students and Teachers program, the Faculty Early Career Development (CAREER) Program, the Albert Einstein Distinguished Educator Program, outreach to minority institutions and entities, and oversees program-level evaluation contracts for research and K–12 education. Dr. Pea's areas of interest include research on professional development, teacher education, teacher beliefs about science teaching, stereotype and identification of threats, and student achievement. Dr. Pea has been involved on the national level with many different organizations, including the National Association for Research in Science Teaching where she serves on the research committee; the National Research Council; the National Association of Biology Teachers; the American Physical Society; and the National Science Teachers Association. She has served as an adjunct professor at George Mason University. She has coauthored several articles in science, has contributed chapters to national and international publications, and has several articles under review. Before coming to the National Science Foundation, Dr. Pea was the science coordinator for a Louisiana statewide reform initiative and a middle school science teacher for East Baton Rouge Parish Schools. She holds the following degrees: bachelor of science, masters of science in biology, and a PhD in science education leadership.

Any opinions, findings, conclusions or recommendations expressed in this chapter are those of the authors and do not necessarily reflect the views of the National Science Foundation in any way.

Part I
Overview

Chapter 1

Introduction to Models and Approaches to STEM Professional Development

Celestine H. Pea and Brenda S. Wojnowski

Identifying a Core Strategy

For more than a half century, school teachers have been both the targets and agents of change for scores of reforms in science, technology, engineering, and mathematics (STEM) education (e.g., Church, Bland, and Church 2010; Cuban 2012; Guskey and Yoon 2009; Porter et al. 2004), all linked to student achievement and the need for the United States to remain at the forefront of discoveries. Today, the focus is no different; teachers remain at the core of strategies for improving teaching and learning in STEM education. This book highlights professional development models and approaches used by several states and districts to significantly improve teaching and learning in one or more areas of STEM education, with significant emphasis on science and, to a lesser degree, mathematics education. The overarching goal for each model or approach included in the book is to develop teachers' and students' knowledge and skills, and, ultimately, to improve student achievement in STEM education.

Although data in 2011 from the National Assessment of Educational Progress (NAEP) show that "scores were higher in 2011 than in 2009 for eighth-grade students at the 10th, 25th, 50th, and 75th percentiles" (NCES 2011, p. 5), overall student performance and achievement remain woefully low, and significant gaps continue to exist between underrepresented students and their counterparts.

To increase overall teacher effectiveness, improve student success, and close the achievement gap nationwide, a variety of professional development models and approaches are underway to move science education along a more positive trajectory for all STEM teachers and students. These efforts are designed around education policies, infrastructure components, and district and state contexts and are geared toward exceeding the NAEP 2011 student achievement results for all students.

The main strategy for professional development models highlighted here stemmed from one or more of the following factors:

- increasing capacity and maximizing improvement through application of local funds or external competitive grants

- engaging in lifelong learning
- implementing new standards or new curriculum, integrating technology and engineering design, or addressing new book adoptions
- garnering political support for and meeting political demands on education
- participating in major STEM education reforms
- improving knowledge accumulation and transfer across multiple STEM audiences and domains
- addressing social issues and responding to economic downturns
- participating in research interventions and related activities

Design Strategy

Many Options, No One Right Answer

The states and school districts featured in this book used state and local data as well as student achievement test scores from national databases, such as those housed at the National Center for Education Statistics (NCES), to inform the development of models or approaches. Efforts also included means to measure the impact of the professional development offered and to better understand what might be causing achievement gaps among subpopulations of students to persist.

Thus, states and districts based their decisions on information from these key assessment measures as they gave careful attention to the design process, procedures, problem analysis, and solutions (Edelson 2002; Fortus et al. 2004) associated with each model and approach. The design process, then, became a powerful tool for demonstrating the interplay between a model or approach and human experiences, knowledge, skills, and new understandings (Figure 1.1, p. 6). Additionally, the design process helped to identify critical leaders; highlight key processes; and underscore ways to analyze and minimize problems in light of local and national goals, needs assessments, and available resources. The models and approaches all required close attention to trade-offs and maximum results while remaining mindful of the challenges, constraints, and opportunities presented by the design process itself (Edelson 2002; Fortus et al. 2004). Therefore, no two models or approaches emerged exactly alike. Although many components are common across the examples presented here, each one is uniquely situated in the context of a local state or school district.

Different Views, Same Goal

Although states and school districts had different views about what their models and approaches should look like, the overarching goal for STEM education was the same: preparing teachers to enable and inspire more students to become productive citizens and to select STEM careers. Each year the National Science Board releases its Science and Engineering Indicators. The Science and Engineering Indicators (NSB 2012) provide evidence that a significantly greater number of students than are currently choosing STEM careers are needed to maintain America's competitive edge in the world marketplace, improve economic stability, and secure national defense. The President's Council of Advisors on Science and Technology (PCAST 2012) and the National Research Council (NRC 2012) are calling for all states and school districts to engage in activities deliberately aimed at reaching, achieving, and maintaining these laudable goals.

In that regard, Chapters 2–4 of this book set the stage for the plans detailed in Chapters 5–12. Chapter 5, "Using Constructivist Principles in Professional Development for STEM Educators: What the Masters Have Helped Us Learn," begins with a summary account of the role of constructivist principles in professional development for STEM educators in a multistate education consortium. Chapter 6, "Ohio's 30 Years of Mathematics and Science Education Reform: Practices, Politics, and Policies," represents a statewide model based on over 30 years of STEM reform and chronicles the confluence of strategies and approaches that flowed from both mathematics and science state agencies along with university partners, local school districts, and external funders and stakeholders.

Chapter 7, "Improving and Sustaining Inquiry-Based Teaching and Learning in South Carolina Middle School Science Programs," features a professional development intervention model aimed at transforming teachers' practice from traditional to inquiry instruction at the middle school level. The intervention focuses on the school as the unit of change, requiring that at least 60% of science teachers agree to participate in the first level of professional development. Chapter 8, "K20: Improving Science Across Oklahoma," looks at an integrated whole-school professional development model that emphasizes leadership and the creation of professional learning communities.

Chapter 9, "The iQuest Professional Development Model," describes a program that brings technology-enhanced learning experiences as early interventions for middle school teachers and students in classrooms with large underrepresented student populations. This focus on students is especially important: The participation of students from underrepresented population is low in STEM fields. Over a two-year period, teachers participate in professional development and mentoring to help them become comfortable with integrating educational tools of technology into science teaching and learning.

Chapter 10, "The Boston Science Initiative: Focus on Science," highlights a school district's struggle to integrate K–12 science into ongoing mathematics and English language arts districtwide reform. This chapter describes the difficulties science leaders faced in convincing district leadership to make such a transition as well as the buy-in by partners and other local education stakeholders whose support made the transition possible. Chapter 11, "Seattle Public Schools' Professional Development Model: Preparing Elementary Teachers for Science Instruction," looks at progressive systemwide change over nearly 20 years. The chapter concentrates on the system and infrastructure developed to support teachers and students at the elementary level. Chapter 12, "New York City STEM Professional Development Partnership Model," describes a progression of strategies, approaches, challenges, and changes inherent to a professional development initiative for providing secondary STEM teachers with the knowledge and skills needed to raise achievement for students enrolled in low socioeconomic status schools. Chapter 13, "Creating and Sustaining Professional Learning Communities," brings the book to a close with a look at professional learning communities as a way for those schools, districts, and states that are working to enhance or reformulate their STEM professional development models to make use of the valuable repositories of information presented in this book. It goes on to describe how this book can be used to develop teacher skills and knowledge in the pursuit of school reculturing for more effective nurturing of teachers and students as they strive for improved STEM achievement.

Figure 1.1

LOGIC MODEL FOR CHANGE AND IMPROVEMENT

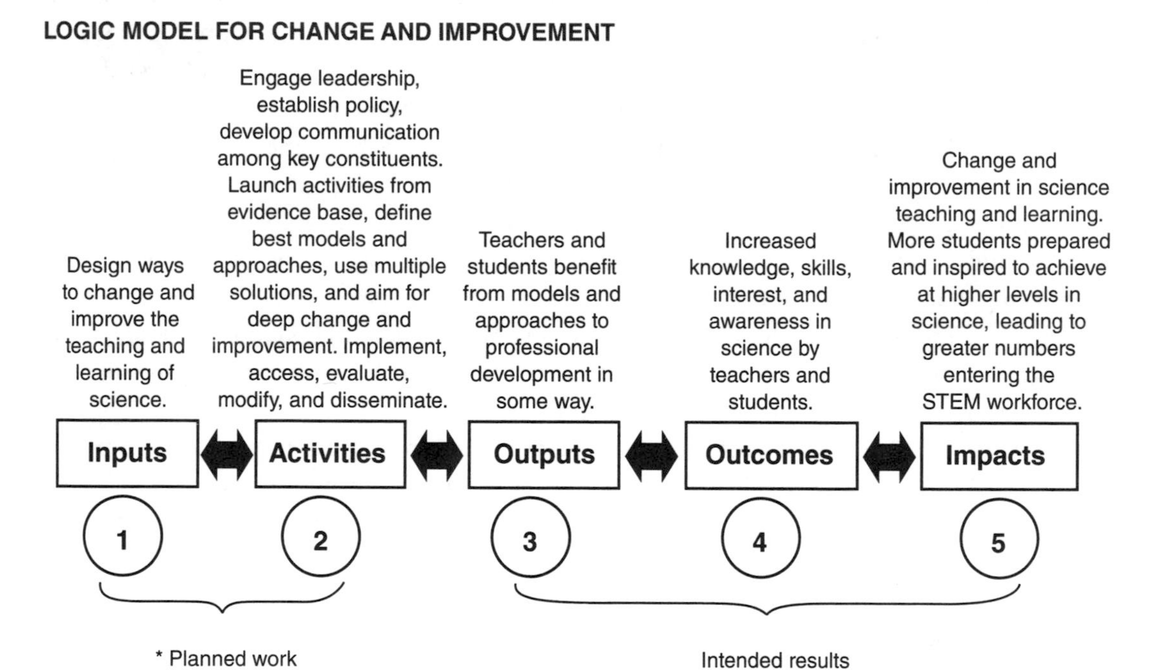

Summary

The models and approaches found within this volume are positioned to share insights about how states and local school districts are contributing to creating a stronger pool of highly effective teachers. The models and approaches describe steps taken to select a focus; identify a unit of change as a target for professional development; outline milestones, key strategies, and decision-making tools and ideas; and share the value of having input from all stakeholders. Although all of the methods focused on STEM teachers, many included administrators and other school leaders as well.

Common across all programs was attention to active learning on the part of teachers and students, content and pedagogy, duration of activities, reform-based strategies, and collaborative practices. Thus, it is evident that enhanced STEM professional learning designed to produce highly effective teachers requires well-designed and well-implemented professional development for teachers. In turn, the expectation is that these teachers will help more students select to enter and complete the requirements for careers in STEM fields (PCAST 2012). These models and approaches might also begin to establish a foundation that can accommodate the resurgence of research about the transfer of knowledge gained by teachers through professional development, to its implementation in formal and informal settings, and to its final use by students in novel everyday life situations and STEM careers (Belenky and Nokes-Malach 2012; Engle 2012; Penuel and Fishman 2012).

Educators, researchers, and policy makers agree that professional development is a complex, expensive, long-term process through which teachers engage in professional learning. Developing and maintaining the capacity of a state or school district to build and sustain behaviors and characteristics needed for improving STEM teaching and learning (Beaver and Weinbaum 2012) rests primarily on the shoulders of teachers. Hence, there will always be reasons for teachers to engage in professional learning—both individually and collectively. We believe you will find this book helpful in your quest for designing and implementing effective models and approaches for enhancing professional learning in districts and states across the country.

Any opinions, findings, conclusions, or recommendations expressed in this chapter are those of the authors and do not necessarily reflect the views of the National Science Foundation in any way.

References

Alsalam, N., L. T. Ogle, G. T. Rogers, and T. M. Smith. 1992. *The condition of education.* NCES 92-096. Washington, DC: U.S. Department of Education, National Center for Education Statistics.

Aud, S., W. Hussar, G. Kena, K. Bianco, L. Frohlich, J. Kemp, and K. Tahan. 2011. *The condition of education 2011.* NCES 2011-033. Washington, DC: U.S. Department of Education, National Center for Education Statistics.

Beaver, J. B., and E. H. Weinbaum. 2012. Measuring school capacity, maximizing school improvement. Policy brief RB-53. Philadelphia, PA: Consortium for Policy Research in Education.

Belenky, D., and T. Nokes-Malach. 2012. Motivation and transfer: The role of mastery-approach goals in preparation of future learning. *Journal of the Learning Sciences* 21 (3): 399–342.

Church, E., P. Bland, and B. Church. 2010. Supporting quality staff development with best-practice aligned policies. *Emporia State Research Studies* 46 (2): 44–47.

Cuban, L. 2012. For each to excel: Standards vs. customization: Finding the right balance. *Educational Leadership* 69 (5): 10–15.

Edelson, D. C. 2002. Design research: What we learn when we engage in design. *Journal of the Learning Sciences* 11 (1): 105–121.

Engle, R. A. 2012. The resurgence of research into transfer: An introduction to the final articles of the transfer strand. *Journal of the Learning Sciences* 21 (3): 347–352.

Fortus, D., C. Dershimer, J. Krajcik, R. Marx, and R. Mamlok-Naaman. 2004. Design-based science and student learning. *Journal of Research in Science Technology* 41 (10): 1081–1110.

Guskey, T. R., and K. S. Yoon. 2009. What works in professional development? *Phi Delta Kappan* 90 (7): 495–500.

Loveless, T. 2011. *The 2010 Brown Center report on American education: How well are American students learning?* Washington, DC: Brookings Institution. *www.brookings.edu/~/media/research/files/reports/2011/2/07%20education%20loveless/0207_education_loveless.pdf*

National Center for Educational Statistics (NCES). 2011. *The nation's report card: National Assessment of Educational Progress (NAEP) at grade 8*. Washington, DC: NCES, Institute of Education Sciences, U.S. Department of Education.

National Research Council (NRC). 2011. *Successful K–12 STEM education: Identifying effective approaches in science, technology, engineering, and mathematics.* Washington, DC: National Academies Press.

National Research Council (NRC). 2012. *Monitoring progress toward successful K–12 STEM education: A nation advancing?* Washington, DC: National Academies Press.

National Science Board (NSB). 2012. Science and engineering indicators 2012. NSB12-01. Arlington, VA: National Science Foundation.

Penuel, W. R., and B. J. Fishman. 2012. Large-scale science education intervention research we can use. *Journal of Research in Science Teaching* 49 (3): 281–304.

Porter, A., B. Birman, M. S. Garat, L. M. Desimone, and K. S. Yoon. 2004. *Effective professional development in mathematics and science: Lessons from evaluation of the Eisenhower Program.* Washington, DC: American Institutes of Research.

President's Council of Advisors on Science and Technology (PCAST). 2010. *Prepare and inspire: K–12 education in science, technology, engineering, and math (STEM) education for America's future. www.whitehouse.gov/sites/default/files/microsites/ostp/pcast-stem-ed-final.pdf.*

President's Council of Advisors on Science and Technology (PCAST). 2012. *Report to the President. Engage to excel: Producing one million additional college graduates with degrees in science, technology, engineering, and mathematics. www.whitehouse.gov/sites/default/files/microsites/ostp/pcast-engage-to-excel-final_2-25-12.pdf.*

Chapter 2

Professional Development: A Historical Summary of Practices and Research

Celestine H. Pea and Brenda S. Wojnowski

Introduction

This chapter provides an historical summary of professional development relative to science education reform. It details an account of how professional development for teachers emerged at the forefront of science education reform and has steadfastly held that position for more than 60 years. Historical records show that, although it took the combined, gradual, but unrelenting, efforts of many scientists and science educators to infuse science into the public school curriculum, early supporters were in general agreement that teachers would need a deep command of subject matter and pedagogical content knowledge to teach science effectively (DeBoer 1991, 2000; Shulman 1987).

In 1947, as the need for science in the school curriculum gained momentum, professional training for teachers took on more prominence. By the mid-1950s, there was general agreement that the quality of science teaching needed to improve substantially (NSB 2000), placing professional development firmly in the center of the scientific movement.

A Policy to Support Professional Development

The National Defense Education Act of 1958 was a policy mandate that grew out of the United States' response to the launching of Sputnik. This act provided support for vocational teacher training and science courses in K–12 schools. However, it was the Elementary and Secondary Education Act (ESEA) of 1965, which included provisions for teacher training in mathematics and science, that further cemented professional development for teachers as the core strategy for helping to prepare the nation's students to become first in the world. Together with the Civil Rights Acts of 1965, the decade following the launching of Sputnik paved the way for foundational changes and improvements in science education.

The Early Years

The 1960s

With an emphasis on improving teaching and learning in science bolstered by funds from ESEA, inservice training in the 1960s was used to denote activities that concentrated on improving teachers' knowledge, skills, and teaching practices. As the strategy for providing opportunities for teachers to improve professionally, inservice training took on many forms in multiple venues, but inservice training primarily occurred after school hours, on student release days, on special workdays prior to the beginning of a new year, or during the summer, with participation often being voluntary.

The content of inservices in those days tended to be generic and may or may not have been geared to a specific discipline, teacher assignment, or grade level. The duration was likely two to three hours in a single day. Typically, teachers convened in the school's auditorium, gymnasium, cafeteria, or library to hear presentations from an external source about topics general to education. With a show-and-tell format, the expectation was that teachers would be able to draw some bits of wisdom from these sessions that could somehow inform their practice. Research on professional development at that time was not very prevalent.

The 1970s

The 1970s brought more structure to professional development. However, it remained mostly voluntary. Again, district- and school-level personnel determined the focus of the courses and workshops offered to teachers. These courses or workshops were either district led or university based and were sometimes provided by representatives from book companies or suppliers of scientific materials and equipment. The importance of these opportunities was measured by top administrators with the expectation that somehow teachers would find some value from these offerings that could be linked back to the classroom. Although the intention was good—good intentions rarely translated into actual or worthwhile changes.

Toward the end of the 1970s, Hall and Loucks (1978; 1979) begin to look at the teachers' voices in the whole professional development construct. Through two studies, Hall and Loucks explored ways to facilitate and personalize professional development to the needs of individual teachers. For example, as new curriculum materials emerged on the science education market, professional development took on the role of helping teachers learn how to use these products to teach science. Recognizing that this could be a problem for school-level administrators and teachers, Hall and Loucks (1979) used the Concerns-Based Adoption Model to study the implementation of the curriculum materials with teachers as the focal point of the adoption process. These researchers also looked at the impact of organizational and social influences on the implementation process. Hall and Loucks noted that the two critical findings from this research were the impact of the principals' perspective on change and how the principals' perspective influenced teachers' views.

The 1980s

In the 1980s, the structure of professional development changed significantly. By that time, states and districts used lessons learned from earlier efforts to build directories of diverse sets of professional development opportunities that were more teacher inspired and content driven. Teachers' voices became even more a part of the process, and the value teachers placed on what was offered became part of the overall consideration for professional development. Further considerations of teachers' opinions led to individual professional development plans and the self-selection of courses and workshops for growth and improvement. Though teachers largely controlled what types of professional development they volunteered to participate in, choices made did not always lead to classroom implementation or map with the types of competencies needed to advance teaching and learning at that time. Additionally, many of the ill-focused inservices of the 1960s and 1970s continued along with the more teacher-driven professional development that was emerging to the forefront.

The 1980s also brought on several shifts in thinking about education as a whole. For example, it was the states' and districts' response to *A Nation at Risk* (NCEE 1983) that ushered in a new decade of educational reform, including an expansion of professional development for teachers around science content, curriculum programs, and instructional strategies. As the result of an urgent plea to reform education for all students, professional development focused more directly on curricular programs along with instructional materials and other teacher-related products that could help instructors address the learning styles and specific needs of a greater number of students.

Perhaps the greatest shift occurred when the Eisenhower Professional Development Program that grew out of the ESEA Act of 1965 to support teachers' learning through high-quality professional development in mathematics and science was reauthorized, amended and reauthorized as the Dwight D. Eisenhower Professional Development Program under Title II, Part B, in 1994. Through additional reauthorizations, the most recent version of the law (USED 2010) focused not only on teachers but on developing highly effective principals as well.

As an outgrowth of this legislation, billions of dollars allocated to states, districts, schools, and universities are leveraged with local monies and support from other sources (e.g., private and public foundations, business, and industry) to provide professional development for teachers (Corcoran, 1995a, 1995b; Cuban 1990; Guskey 1994; Porter et al. 2004; Spillane and Callahan 2000). These partner-efforts became a mainstay for professional development even as science reforms shared the national interest spotlight with space exploration programs (NSB 2000).

The combined efforts of national and local policy, proliferations of professional development, improvements in social conditions, and the upturn in economic stability led to substantial advancements in student achievement between 1971 and 1988 (O'Day and Smith 1993). The dramatic gains for underrepresented minority students that likely occurred from the impact of the Civil Rights Act of 1965, teachers' ability to teach per the knowledge and skills requirements at that time, and ESEA closed the achievement gap by 30–60 points, depending on the grade assessed. Unfortunately, toward the end of the 1980s, social and economic conditions took a downturn and education gains and student achievement quickly followed (O'Day and Smith 1993).

Research about many factors related to teachers and models that centered on teaching, professional development, and adult education began to flourish during this decade (Joyce and Weil 1980; Knowles 1980; Sparks and Loucks-Horsley 1989). Other studies targeted elements of teaching in which the focus was teachers and change (Guskey 1982; 1984; 1986; 1989), peers and colleagues (Little 1987), teachers' thinking as related to teaching, and teachers' knowledge creation and use in teaching (Elbaz 1983; Eraut 1988). Still other studies, books, and reports examined teachers' pedagogy and pedagogical expertise (Berliner 1988; Shulman 1987) and the involvement of higher education in teacher practice (Bloom 1988). Even further studies looked at teachers' skills and at their personal and professional identities (Huberman 1985; Nias 1989).

Data from a national survey conducted in 1985 showed that teachers ranked inservice training as highly ineffective in helping them learn and grow in the teaching profession (Smylie 1989). A review of the literature on professional development by Guskey (1986) found that nearly all of the research on existing professional development efforts criticized the effectiveness of what was being offered. Despite those discouraging assessments, professional development remained the best strategy for improving learning at the time (Huberman 1995). As the decade came to a close, Rosebery, Warren, and Conant (1989, 1992) studied how best to engage students in science when English was the second language. This research brought culture into the professional development arena in ways not evidenced before.

One of the earlier methods for looking at teacher and student thinking was cognitively guided instruction (CGI), designed to help make teachers' knowledge about how students think, particularly in mathematics, more formal, organizationally sound, and coherent. This method launched a whole new movement in professional development around students' thinking as subsequent generations of CGI evolved. A review of the literature in mathematics and science (Kennedy 1998) showed that CGI was one of the first projects that linked studies of teacher learning and knowledge to student achievement.

A New Perspective for Professional Development

The 1990s

The 1990s brought a series of twists and turns that changed the face of professional development dramatically. Recognizing that student gains that grew out of the 1960s and 1970s were declining or virtually flat, educators and researchers began to look at other solutions that might help ameliorate the problem (NSB 2000; Smith and O'Day 1991). Educators, policy makers, and researchers began to look at professional development not as a single activity isolated from other school-related events. Instead, these stakeholders realized that to meet the needs of all students and teachers, the unit of change needed to shift from the teacher to a whole school, district, state, or region.

Therefore, as the decade continued, science education was guided by the systemic reform theory to a significant degree, and many states and districts moved to system thinking (Senge 1990). The hypothesis was that in order to bring about enduring reform, attention should be paid to all parts of the system (e.g., standards, policy, curriculum, fiscal, partners, and achievement) simultaneously. This movement marked the first time that a clearly expressed theory served as the linchpin to reform under the premise of "complex solutions to complex problems" (NSB 2000, p. 5–8).

Though untested in educational settings, the theory teetered precariously on the principle that "there are too many complex, interconnected problems present for any one, simple isolated solution to alter the fundamental dynamics of teaching and learning in the overall education system or even a single classroom" (NSB 2000, p. 5–8). The systemic reform theory, then, attempted to reign in isolated ideas about reforming science, technology, engineering, and mathematics (STEM) education. Thus, professional development again became the variable that led to the amalgamation of funds from ESEA, public and private foundations, and business and industries in support of a comprehensive systemwide strategy for reforming STEM education (see Figure 2.1; Cuban 1990; Guskey 1994; Marx et al. 2004; O'Day and Smith 1993; Yin 2006).

Figure 2.1

SAMPLE SOURCES OF SUPPORT FOR TEACHER-DIRECTED STEM ACTIVITIES

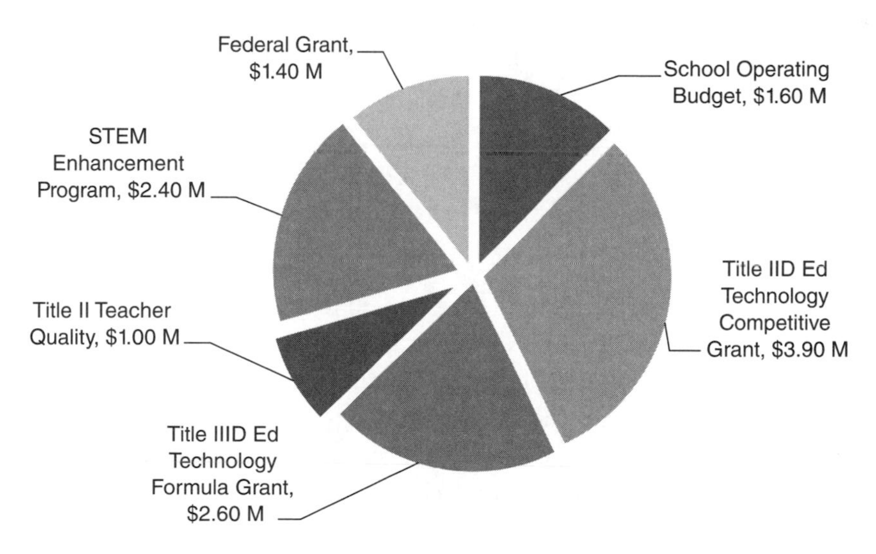

In the push for systemwide change, professional development maintained its place as the trademark for science education reform for more than a decade (Corcoran, Shields, and Zucker 1998; Guskey 2009; Yin 2006). Though much of the attention during the 1990s was on systemic reform, many states and districts continued to use professional development as the core strategy for reforming education. With support from major partners, states, rural partnerships, and urban school districts promoted professional development through discipline-focused summer institutes, yearlong courses, afterschool and weekend workshops, and other venues in which teachers and students became the centerpieces for changing science education.

At that point, many state and local officials and professional developers realized that single, or even multiple teacher professional development activities could not provide all that was needed to meet the demands for ensuring that all students had access to highly qualified, highly effective teachers capable of teaching science in ways that impacted student performance and achievement along a positive trajectory. This acknowledgement led to increased use of the train-the-trainer model for professional development. With this approach, the systemic initiatives alone supported hundreds of teacher leaders in their leader roles from the early 1990s to midway into the first decade of the 21st century (Yin 2006). Many of these teachers were either partially or fully released from their classroom duties to support other teachers in improving science instruction in their classrooms. Together, at the school and district levels, teacher leaders provided thousands of additional hours of professional development and training for classroom teachers, administrators, and parents specific to the role each played in the teaching and learning of science.

In addition, through the combined efforts of university-based courses in science content and pedagogy, summer institutes and workshops, weekend and afterschool events, internships, externships, and specialized programs designed specifically for content mastery and advanced degrees (e.g., masters in science), teachers had access to many opportunities that promoted professional and personal growth. With increased accountability for improving science education and high-stakes testing, states and districts moved from offering mostly voluntary opportunities to those that were required (e. g., new curriculum, new policies), were specific to teaching assignments, or were tied to district credits, recertification requirements, or other policy mandates.

The Onset of Technology

During the 1990s, many different strategies were used to reach as many teachers as possible to ensure that they obtained the knowledge and skills needed to improve teaching and learning nationally and locally. Despite professional development being commonplace in many settings and environments, the use of technology through online and other learning environments surfaced as a major equalizer for teachers who lived in rural and isolated areas, faced problems with childcare, or lacked the time to participate in professional development offered solely through face-to-face venues. As technology becomes more prevalent and discoveries in its use in education more readily available, more teachers and students gained access to and became part of the worldwide connected community of learners (Bos, Krajcik, and Patrick 1995; Casey 1994; Edelson, Gordin, and Pea 1999). New reforms recommended the use of networking technologies in K–12 schools to help teachers change from teacher-centered to student-centered practices (Darling-Hammond and McLaughlin 1995) and to lower the cost for helping teachers to engage in lifelong learning.

Schools and districts found it useful to capitalize on the availability of technology to enhance both online and face-to-face professional development opportunities. Through this blended approach, states and districts could better build the school's or district's capacity and keep the expertise within the respective learning communities. While providers external to a system are important to any professional development model or approach, building a system's capacity to move learning forward should be of primary importance (Porter et al. 2004; Yin 2006). Engaging in a blended approach allowed schools and districts to provide professional development at times not possible otherwise and to embed opportunities in the school day so that teachers could practice new strategies with their own students in their own classrooms.

With the rapid expansion of online learning, many schools looked for innovative ways to use technology in "small doses" to provide professional development that could be easily available on a regular basis. A variety of technology-based short tutorials helped teachers obtain additional professional development through special channels that offered free daily skills for teachers and students. For example, topics might focus on how to incorporate Google Earth into a lesson, develop assessment in a digital format, use magic ink, or look at "live" cell division or the amoeboid movement of microorganisms. Nevertheless, not all educators bought into the bite-sized chunks but remained stout supporters of face-to-face professional development. Those who preferred face-to-face approaches found this method was more engaging, allowed for collegial interactions, and offered tailoring of activities for specific training (see *www.edweek.org/ew/articles/2011*).

Today, social networking, digital webinars, online courses, digital storytelling, blogs, podcasts, and Skype are just a few of the ways that states, districts, schools, and individuals continue their lifelong learning through technology. Researchers are also beginning to explore how social networking adds to the venue through which teachers grow professionally. To assist schools with advancing the use of technology in professional development, scores of technology grants have helped states and districts change the culture of their school to a more collaborative one. Under this mantra, technology became more than a skill-drill tool or a presentation device and was seen as a way to improve instructional practice and engage students.

The Arrival of Science Standards

As the systemic reform movement took hold, challenging instruction such as hands-on and inquiry-based methods began to emerge as part of the core for developing a more scientifically literate citizenry. The American Association for the Advancement of Science's *Science for All Americans* (AAAS 1989) and *Benchmarks for Science Literacy* (AAAS 1993) and the National Research Council's *National Science Education Standards* (NSES; NRC 1996) were all at the leading edge of producing a new set of standards for science education (Loucks-Horsley and Bybee 1998). These documents helped promote a universal understanding of what good teaching and learning should entail. Research-based books (e.g., Loucks-Horsley et al. 1998, 2003, 2009) helped narrow ideas and refine the roadmap educators would use in planning effective professional development for teachers.

During this time frame, professional development concentrated largely on the NSES standards for professional development and the changes the Standards brought to teaching through inquiry instruction. Regardless of the nature of the providers, face time became of longer duration, the focus on content stressed "less is more," and the design strategy became more student oriented.

Despite the standards, groundbreaking policies (e.g., the Goal 2000: Educate America Act of 1994), significant funding, theory-driven reform, and NSES, student achievement in science remained woefully low (Loveless 2011). Evidence from evaluation studies (e.g., Porter et al. 2004) and more frequent use of data and results from the National Center of Education Statistics (NCES 2009, 2011) and the Third International Mathematics and Science Study (NCES 2001), as well as reports from Learning Forward (*www.learningforward.org*), showed that many teachers and students were not getting the support they needed.

A Growing Research Base

There was an explosion of research about every aspect of professional development imaginable during the 1990s. A glimpse into the evidentiary base from which researchers and expert practitioners assisted states and school districts can be gleaned from contributors such as Fullan and Steigelbauer (1991) who explored the meaning of educational change. Fullan and Miles (1992) noted that one of the underlying reasons that professional development was not as effective as expected was the lack of attention to what it actually requires for change to occur in educational settings. Murphy (1992) looked at the role of study groups in fostering schoolwide learning; Cuban (1992) reported on reforms that last despite the hardships that often challenge and impede change, and McLaughlin (1991) investigated what had been learned about professional development early on in the decade. Rosebery, Warren, and Conant (1992) searched for new methods that could be used to document teachers, learning as a result of their participation in professional development. One strategy involved looking at the discourse among groups of teachers from different backgrounds and expertise and the influence of those factors on the group's shared meanings of what professional development was attempting to do.

With an increase in the diversity of the public school student population, bilingual education posed new issues for providers of professional development. Rosebery, Warren, and Conant (1992) found that students' understanding and language around scientific thinking deepened dramatically along with their science knowledge concerning the specific topics they studied. Professional development for teachers in this setting was through collaborative inquiry in which teachers guided students' involvement while learning how to use these instructional methods in their own classrooms.

Professional communities and networks for teachers grew in popularity, took on many forms, and offered teachers and educators multiple avenues for professional growth (Lieberman and McLaughlin 1992; Pennell and Firestone 1996). Through these avenues teacher professional development moved beyond geographical boundaries and nested situations to learning across a particular domain or environment.

O'Day and Smith (1993) and Corcoran and Goertz (1995) reported that there must be strong connections among all reform strategies geared toward school change. According to Fullan (1993) and Lieberman (1995), there are key factors that support teacher learning. Primary among the key factors that consistently show up in school improvement research is the close relationship between professional development and school change (Fullan and Steigelbauer 1991). Just as Yin (2006) later reported on the critical role of the superintendent in district change, Marsh and LeFever (1997) found that school leadership is also critical to school reform.

Perhaps one of the most notable contributors to research in the 1990s was Guskey (1994), who advocated for the inclusion of evaluation in the professional development process. To help researchers and educators address the need for adding evaluation, Guskey developed a five-point model for evaluating the impact of professional development on teachers and students. Guskey described the five levels as follows: (1) participant reaction; (2) participant learning; (3) organizational support and learning; (4) participant use of new knowledge and skills; and (5) student learning outcomes.

Further work by Lord (1994) and Bailey and Bailey (1995) looked at the critical role of colleagueship regarding teachers' interactions with one another. Newmann (1994) contributed to the body of research at that time by studying professional communities and school restructuring to meet the need of a more diverse body of student learners. Cook and Rasmussen (1994) addressed the change-based inquiry process by developing a framework for designing professional development, while Fine (1994) studied ways to redesign professional roles and relationships, and Guskey (1994) discussed ways to provide the best catalog of options for ensuring individual, organizational, and student change.

For Parke and Coble (1997), curriculum development became the venue for professional development and school reform involving teachers, students, and administrators. They developed a five-phase model that linked theory and practice through curriculum decision making for middle grade science teachers. Results showed that this professional development design engaged all participants in the decision-making process, which embodied ongoing testing, revising, and reevaluating of curriculum and instructional factors. Throughout the decade, some teachers made remarkable gains in pursuing scientific discourse, but many more struggled to adopt a scientific language and to own the concepts, ideas, and inquiries offered through professional development.

Professional Development: A New Era for Improvement

Into the 2000s

At the dawn of the 21st century, professional development continued to be a multifaceted phenomenon that had fundamentally transitioned from a teacher-focused to a student-focused enterprise. About the same time, the frequency of large-scale professional development prevalent during the 1990s began to wane, primarily due to ending of the systemic initiatives that supplied thousands of hours of professional development for science and mathematics teachers and the enactment of new national education policies. Many positions that supported teacher development and change in classroom practices were eliminated, and the teacher leaders, content specialists, coaches, and teachers on special assignments who were providers of professional development in their schools' districts, or states, found themselves returning to the classroom or moving up the leadership ladder. Science education took a major downturn and lost significant grounding in the school curriculum. To the degree possible, and with support from partners, states and districts continued to offer some form of professional development for science teachers. Whereas before a significant number of science teachers had participated in four to six weeks of activities through 6-hour days, the average duration declined to 15 to 40 contact hours spread over one to six weeks, with most science teachers receiving far less.

However, the conditions were not all bad. Differentiated instruction, constructivists' approaches, and other strategies found to be effective in professional development served as rich areas for practice and research. Study groups, teacher networks, mentoring relationships, committees or task forces, internships, individual research projects, or teacher research centers, were all shown to be better options than traditional workshops or conferences. Unlike the earlier inservice activities that were not teacher or student directed, discipline-specific activities based on teacher assignments or particular science concepts were found to be most useful in the long term (Porter et al. 2004; Wei, Darling-Hammond, and Adamson 2009).

As in the 1990s, states and districts placed greater emphasis on learning communities, whole-school change, or the collective participation of groups of teachers and administrators from the same district, school, department, or grade level, as opposed to the participation of individual teachers from many schools. Lone reformers were not found to be highly effective at any level. The extent to which professional development providers offered opportunities for teachers to engage in active learning was more evident during this time than in prior years. Under this scenario, active learning took on a dual approach aimed at benefiting teachers and their students. For the teacher, this meant having opportunities to become actively engaged in the meaningful analysis of all aspects of teaching and learning from participation, reflection, and modification to change in practice. It also meant having opportunities to actively engage in activities that mirrored what teachers were expected to do with their students in their own classrooms as well as opportunities to review student work or obtain feedback on their teaching. Finally, professional development focused on the degree to which the activity promoted coherence in teachers' professional development by encouraging continued professional communication among teachers and by incorporating experiences that are consistent with teachers' goals and aligned with state standards and assessments (Porter et al. 2004).

As knowledge about and use of technology became more ingrained in the fabric of life and education, technology again became a tool for providing professional development, as well as for reducing teacher isolation, reaching teachers in remote areas, fostering reflection on practice, influencing teaching practices, and supporting the development of professional communities of practice. The evolution of technology changed from just a source for the exchange of information to a recourse that allowed teachers to engage in almost every aspect of learning as part of an evolving community (Corcoran, Shields, and Zucker 1998).

Teacher participation in professional development embedded in research began to become more common. Researchers were interested in exploring why past reforms had not led to major changes in classroom teaching practices, increased student achievement, closed the achievement gap, or increased the number of students electing to pursue careers in STEM fields.

More Research Linked to Practice and Culture

Research on, about, and as a context for professional development advanced to epic proportions during the first decade of the 21st century. Rich examples of research on professional development practices encompassed topics ranging from becoming more culturally competent (Gay 2002) to the use of technology to improve teaching and learning.

Supovitz and Turner (2000) showed that the cumulative results from years of practice underscored what a vision for professional development should entail. Supovitz and Turner found that the quantity of professional development is strongly linked with inquiry-based teaching practice, an investigative classroom culture, teachers' content preparation, school socioeconomic status, principal support, and available resources. Feldman (2000) expanded the argument about conditions that need to be met for teachers to change the theories that guide their teaching practices. Guskey (2002) synthesized information from years of reform by practitioners, researchers, national education organizations, and other education stakeholders to compile a list of elements that best characterized effective professional development.

Soon other researchers (e.g., Gay 2002) advocated for culturally relevant practices to be included in professional development to help teachers become more aware of how to fully meet students' academic needs. From a constructivist perspective, Bryan (2003) examined the belief systems of prospective elementary teachers about science teaching and learning and documented the teachers' beliefs about how children learn science and the students' and teachers' roles in learning science. In this case, the teachers' dualistic beliefs formed two contradictory nests in which the first held an orientation toward didactical practice based on prior experiences and the second embraced a hands-on approach that guided the vision of practice based on current research. The findings accentuate the complexity and nestedness of teachers' belief systems and underscore the significance of identifying prospective teachers' beliefs early in teacher preparation and professional development program activities.

Marx et al. (2004) conducted a three-year study that challenged teachers and students in a large urban school district to collaboratively engage in inquiry-based and technology-infused curriculum units developed by district and university personnel. The initial results were statistically significant for curriculum-based test scores for each year of participation, evidenced by increasing effect size estimates across the years. The findings show that historically low performers in science succeed in inquiry-based science when there is alignment among curriculum, teacher professional development, and district policies. Schneider, Krajcik, and Blumenfeld (2005) investigated how using innovations linked to curricular material might help teachers enact the reforms embedded in the materials.

Lee et al. (2004) were among the researchers who promoted the study of science and literacy for culturally and linguistically diverse elementary students. These researchers examined the impact professional development intervention through instructional units and teacher workshops had on teachers' beliefs and practices related to inquiry-based science. Johnson, Kahle, and Fargo (2007) conducted a longitudinal study of middle school science teachers to investigate whether or not a relationship exists between teacher participation in professional development and student achievement in science. From a different perspective, Corcoran (2007) reported on how policy makers can further improve teachers' knowledge and skills leading to higher-quality teachers and teaching, while Metz (2008) discussed the roles teachers play in educational reforms. The push for clarity about educational change and the role of and impact on teachers continued through reports from noted educators (e.g., Davis 2002; Fullan 2007a, 2007b). Akerson and Hanuscin (2007) conducted such a study over a three-year period, during which teachers participated in a project that emphasized scientific inquiry and inquiry-based instruction. The goal was to help teachers improve their own perspective of the nature of science.

The use of distance education has increased exponentially over the last few decades by institutions of all types and at all levels of education. Annetta and Shymansky (2006) studied the relative effectiveness of three distance education strategies for enhancing the science learning of elementary school teachers who participated in a five-year professional development project. Teachers were engaged through live, real-time interactive television; interactive television with live discussions wrapped around videotaped presentations; and asynchronous, web-based sessions with streamed videotaped presentations supported by interaction through discussion boards. Results showed that participants who engaged via the live mode outperformed participants in the web and video modes.

A rich body of research also emerged on professional development relative to preservice and inservice teachers. For example, Bianchini and Cavazos (2007) followed preservice physical science teachers through their first year of teaching to investigate their efforts to learn to teach science through the ethnic, sexual, linguistic, and academic diversity of their students. In another study, Schneider, Krajcik, and Blumenfeld (2005) reported that inservice teachers face challenges with content and with responding to students' ideas, pedagogy, and willingness to change their approach to teaching science. In contrast, research on preservice teachers by Schwartz et al. (2008) showed that the challenges teachers faced centered on emerging personal beliefs and perspectives, meanings hidden in the language used in curriculum materials, and alignment of review criteria with their own goals. These differences in challenges concerning curricular reform highlight how states and districts might transition to a new way of thinking about professional development for preservice and inservice teachers.

Several multiyear evaluations of the Eisenhower Professional Development Program concluded that professional development in the United States falls short of expectations (Porter et al. 2004; Wei et al. 2009). Further evaluation reports by Porter et al. noted that district-level teachers spent on average 25 hours in ESEA-assisted professional development activities (which had actually doubled from 1988–1989). However, for teachers in higher education–led activities, the time spent was more than double again at 51 hours and with a longer time span than district activities. Whereas 46% of teachers in the university-led activities lasted at least 6 months (including 2% lasting more than 1 year, 20% lasting 10–12 months, and 24% lasting 6 to 9 months), only 20% of district teachers participated in activities that lasted at least 6 months. The performance indicator for ESEA for sustained professional development requires that at least 35% of teachers in a particular school be in activities that extend over the school year. Overall research results showed that district-led professional development activities often do not yet meet the standard, whereas many university-led professional development activities exceed the standard by a substantial amount.

From 2010 to the Present

Currently, professional development in science continues to be composed of a mixture of inservice workshops, summer institutes, workshops, study groups, specially designed science courses, internships, and specialized programs. As was true in the 1950s and 1960s, the driving force behind new policies and requirements is the push for more students (and the entire citizenry) to become more scientifically literate and to select careers in a STEM field.

For those reasons, teachers remain the agents and targets of reform (Metz 2008). To assist teachers in meeting the elusive goal of helping students achieve to higher standards, the reauthorization of ESEA (USED 2010) shifted the focus from being solely on teachers being identified as "highly effective" to including provisions for principals to become "highly effective" as well. ESEA prioritized principal effectiveness to elevate a great school leader in every school to the same level of importance as a great teacher in every classroom. Specifically, ESEA notes the following:

We will elevate the teaching profession to focus on recognizing, encouraging, and rewarding excellence. We are calling on states and districts to develop and implement systems of teacher and principal evaluation and support, and to identify effective and highly effective teachers and principals on the basis of student growth and other factors. These systems will inform professional development and help teachers and principals improve student learning. In addition, a new program will support ambitious efforts to recruit, place, reward, retain, and promote effective teachers and principals and enhance the profession of teaching. (USED 2010, p. 4) The proposed changes installed a new lens for professional development. To ensure that new requirements proposed by the blueprint and other reform measures such as the *Next Generation Science Standards* (*NGSS*; NGSS Lead States 2013) are introduced to teachers and principals, funding has been made available from a variety of public and private sources for professional development that meets the characteristics of these new strategies for ensuring highly effective professional learning.

To help meet the laudable goals of these latest reform efforts, professional development continues to be offered with a clear focus on fewer fundamental and foundational concepts that promote better learning for more teachers, principals, and students. However, the duration of current professional development offerings is not commensurate with what research and expert practitioners identify as highly effective in bringing about classroom change. Instead of the 80 contact hours cited as being necessary to bring about change in the classroom, most teachers actually participate in far less. Most striking about the fewer hours is that a significant percent of the offerings are traditional as well. This condition represents a dichotomy in thinking and raises questions about whether teachers are being adequately prepared to meet newer, higher standards. This is particularly true for the *NGSS*, which set the bar for future teacher professional development at a much higher level than prior efforts. While educators, scientists, researchers, and other stakeholders see *A Framework for K–12 Science Education* (NRC 2012) and the *NGSS* as potentially providing more support for science in the school curriculum at all grade levels, dissenting voices are invoking doubt that this will be the case.

Research Becomes Deeper and More Far-Reaching

With ongoing broad-based policies about professional development, there is a growing commitment to the development of new teacher performance assessments to evaluate what teachers actually do in the classroom. This task is being overseen by a group of states that are developing a common assessment for beginning teachers (Darling-Hammond 2010). Crowe (2010) noted that new assessments should indicate whether teachers have mastered the classroom teaching skills needed to be effective with students who are more diverse and come from more cultures than ever before (Trumbull and Pacheco 2005). Although Crowe suggests that it might be useful to include teachers in examining student gains as part of their overall evaluation, most states have yet to endorse this idea.

Kober and Usher (2012) and other data sources (e.g., Ingersoll and Merrill 2010; Ingersoll 2011) show that the teaching workforce in public schools is rapidly becoming more female (see Figure 2.2) and white, whereas the student population is becoming more diverse.

In addition to never having a male teacher, many students may not experience having a teacher of color during their K–12 school years (see Figure 2.3).

Figure 2.2

TEACHER POPULATION: CHANGE IN SEX 1986–2011

Despite various efforts to increase the male teaching population, including alternate routes to teacher certification, the percentage of male teachers in the teaching workforce continues to decline.

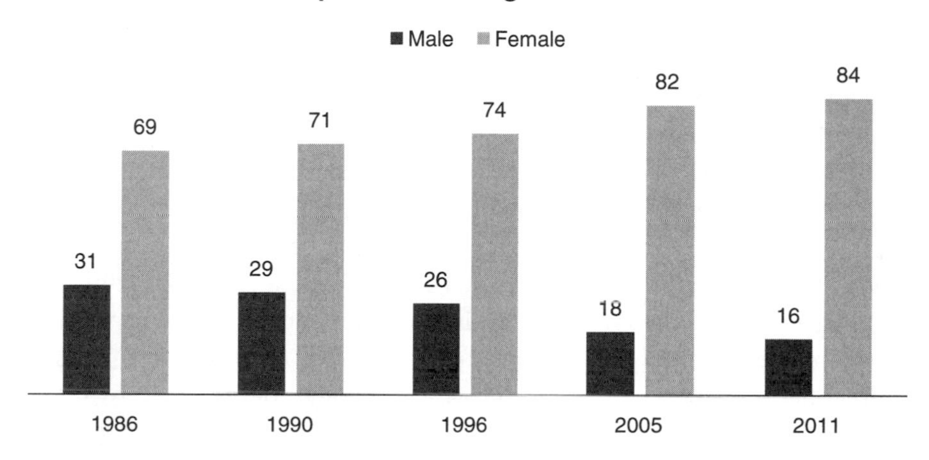

Source: Kober and Usher 2012.

Figure 2.3

DIVERSITY OF THE U.S. TEACHING POPULATION

Despite an overall decline in the diversity of the U.S. teaching workforce, there is a slight increase in the number of Hispanic teachers, 22% of whom are male.

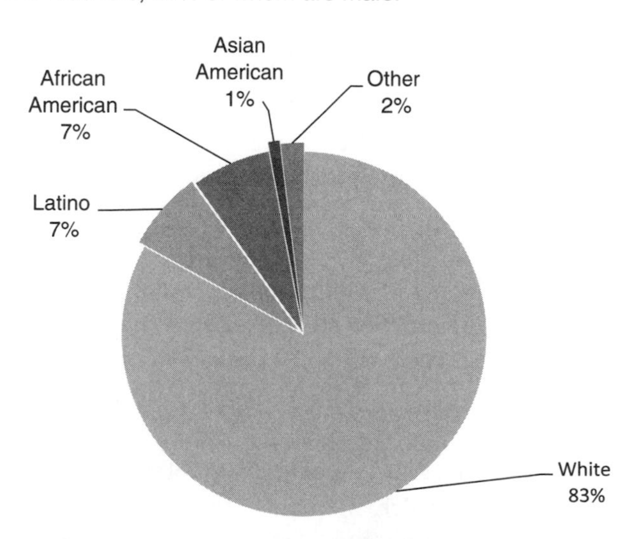

Diversity of the U. S. Teaching Population

Source: Kober and Usher 2012.

Ingersoll (2011) sees this as a major problem given the roles (e.g., role model, surrogate parent, mentor) some teachers play in the lives of their students. While assuring that attention to culture and diversity is important, researchers and educators also acknowledge that doing so adds to an already overburdened professional development strategy.

Roth et al. (2011) analyzed videos of professional development practices to improve teacher and student learning at the upper elementary level. Basing their work on the lenses of a science content story line and student thinking, over the course of one year the researchers showed that the project significantly improved teachers' science content knowledge and ability to analyze science teaching. Similarly, the students increased their science content knowledge.

Researchers and educators are using many different approaches to involve teachers in collaborative work, including inclusive classroom studies (Mutch-Jones, Puttick, and Minner 2012). Using a modification of the Japanese Lesson Study, these researchers created a project for middle school teams made up of both science and special education teachers who engaged in collaborative work to improve instruction in inclusive classrooms. Results from this intervention suggest that science and special educators who participated in the intervention generated more accommodations for students with learning disabilities. Additionally, teachers also increased their ability to develop engaging instructional context and to adapt instructional plans to meet science learning goals for all students in an inclusive classroom. However, despite these gains, the participants did not show significant increases in science content knowledge or content about learning disabilities.

Many researchers and educators believe that learning progressions, or representations of how teacher and student ideas develop in a domain, hold promise as tools to help teachers embed formative assessment into their practices. As such, Furtak (2012) reported on using a learning progression for natural selection to support teachers' enactment of formative assessment. Results indicated that during assessment conversations, teachers could relate students' ideas to parts of the learning progression, and a few teachers used learning progressions to simply catalog misconceptions rather than as developmental affordances.

While Earley and Porritt (forthcoming) used models to evaluate professional development, many researchers and educators still struggle with finding ways to demonstrate that professional development is effective. Others point to studies involving professional development that show the value of considering both teachers and students in the learning process. For example, Maskiewicz and Winters (2012), documented the importance of focusing on students and teachers as they examined teacher and student change through a learning progression while the teachers were engaged in a STEM-based professional development study on inquiry teaching.

Fields et al. (2012) illustrated that middle and high school students scored better on science subject tests when their teachers participated in STEM-based professional development in the same science subject. Studies such as these give further support to the assumption that professional development should focus on both teachers and students when establishing a baseline from which to measure change.

A study by Grimberg and Gummer (2013) reviewed a professional development program for science teachers that looked at culture from a tribal, science teaching, and science perspective. The researchers examined the impact of the program on teachers' practice and beliefs to better understand the influence of and relationship between culturally relevant pedagogical instruction and student gains on science tests. While changes in beliefs or teaching strategies of comparison teachers were not evident, after two years in the program, teachers who participated in the intervention changed their teaching practices and beliefs regarding their ability to teach science equitably in ways that positively impacted students' performance.

Herold (2013) of *Education Week* reported on a team of researchers led by Barry Fishman, an Associate Professor at the University of Michigan, who investigated whether online professional development has the same effect on student learning and teacher behaviors as traditional face-to-face approaches. To test which methods might be more effective, Fishman compared the outcomes of a group of teachers who participated in 48 hours of face-to-face workshops spread over six days to a group of teachers who accessed the same information online at their own pace. The researchers found no differences in the degree to which teachers in both groups reported confidence in their ability to implement the curriculum as prescribed by the designers of the high school curriculum on environmental science. Beyond the benefit to both teacher groups was the evenness of achievement by students of both groups.

A Promising Future

New Approach by and for Diverse Stakeholders

Despite decades of long-term support and less than large-scale improvements in science education as a whole, scores of partners continue to back professional development as the single most critical factor in reforming science education. Through the blueprint (USED 2010), funds that targeted K–12 education reached $3.2 billion in 2011. Of that amount, $2.9 billion were set aside to prepare and recruit teachers and principals, while a small portion ($180 million) targeted STEM formula-driven partnerships and educational technology state grants (USED 2010; PCAST 2010, 2012).

A two-pronged approach (1) to prepare all students so they have a strong foundation in STEM no matter what careers they pursue and (2) to inspire students so that all are motivated to learn STEM subjects underlies this massive investment and guides core elements of the *Blueprint for Reform* (2010). One of the key features of the government's approach,called 100Kin10, is to recruit and train 100,000 highly effective STEM teachers over the next decade to prepare and inspire students. To reach this pinnacle, teachers will need deep content knowledge in STEM subjects, culturally relevant resources and materials, and mastery of the culturally relevant pedagogical skills required to teach STEM subjects well (PCAST 2012).

The Carnegie Foundation of New York (2009), together with the Institute for Advanced Study, is holding steadfast in support of educational reform. Working with the U.S. Department of Education and garnering support for 100Kin10, the call is for upping the professional development of thousands of highly effective STEM teachers in five years and to invest heavily in STEM teacher preparation as well. As noted earlier, the call for better prepared teachers is always linked to addressing some parallel need for students. This is just one example of a broad-based effort aimed at doing just that and with the expectation of producing 1,000,000 more STEM graduates by 2022 (PCAST 2010, 2012).

Summary

This brief historical account traces the role of and support for teachers through inservice activities in which each change in reform efforts and terminology emphasizes the deeper importance of providing for teachers, principals, and students. From the 1940s when the focus was on what science education in the school curriculum should be to the *NGSS*, which continue to grapple with this question, there is universal agreement that any call for improvement in student achievement will result in a parallel call for professional development for teachers. The move forward from teacher training in the 1950s to the present calls for professional development that makes it clear that what is needed far exceeds mere informational knowledge. Instead, science education has moved much closer to the need for being able to translate what is known into models and approaches that transform science learning.

Over the years, many models and approaches have been developed and abandoned based on the demands and sophistication of each new reform. Records show that although many states and school districts responded to each new shift in direction, a host of school-related factors (e.g., economic, educational, political) made it almost impossible to bridge the gap between existing policies and practices and new ones. Longitudinal district and student data reveal that changing teaching practice and improving student performance is the primary role of professional development. Evidence also shows that students whose teachers participate in professional development or lifelong learning achieve at higher levels than students whose teachers do not. This evidence alone further verifies that all teachers need opportunities to grow professionally and continually.

Teaching is a profession that requires continuous professional development. However, no one can pinpoint where and when all professional learning takes place and under what conditions. What is assumed is that with the onset of the knowledge age and explosions in technology, the available sources from which teachers can draw information are boundless. Often such information comes from experiences in formal and informal environments; methods and foundations courses in education and the arts and sciences; fieldwork; online resources, books, journals, and magazines; and every school experience in which teachers engage, all of which have the potential for contributing to teacher professional growth.

Any opinions, findings, conclusions, or recommendations expressed in this chapter are those of the authors and do not necessarily reflect the views of the National Science Foundation in any way.

References

100Kin10. 2012. Answering the nation's STEM challenge. *100kin10.org*

Akerson, V. L., and D. L. Hanuscin. 2007. Teaching nature of science through inquiry: Results of a 3-year professional development program. *Journal of Research in Science Teaching* 44 (5): 653–680.

American Association for the Advancement of Science (AAAS). 1989. *Science for all Americans*. New York: Oxford University Press.

American Association for the Advancement of Science (AAAS). 1993. *Benchmarks for science literacy*. New York: Oxford University Press.

Annetta, L. A., and J. A. Shymansky. 2006. Investigating science learning for rural elementary school teachers in a professional-development project through three distance-education strategies. *Journal of Research in Science Teaching* 43 (10): 1019–1039.

Bailey, D., and S. Bailey. Aligning community: More-than-rational approaches to whole-system change. Workshop presented at the conference of the National Staff Development Council, 9–13 December 1995, Chicago, Illinois.

Berliner, D. C. 1988. Implications of studies on expertise in pedagogy for teacher education and evaluation. In *New directions for teacher assessment*, ed. J. Pfleiderer, 39–68. Princeton: Educational Testing Service.

Bianchini, J. A., and L. M. Cavazos. 2007. Learning from students, inquiry into practice, and participation in professional communities: Beginning teachers' uneven progress toward equitable science teaching. *Journal of Research in Science Teaching* 44 (4): 586–612.

Bloom, D. 1988. The role of higher education in fostering the personal development of teachers. In *Teacher education and the world of work: New economic, social and professional imperatives for the twenty-first century*, ed. H. Hooghoff and A. M. van der Dussen, 59–71. Enschede, the Netherlands: National Institute for Curriculum Development.

Bos, N. D., J. Krajcik, and H. Patrick. 1995. Telecommunications for teachers: Supporting reflection and collaboration among teaching professionals. *Journal of Computers in Mathematics and Science Teaching* 14 (1-2): 187–202.

Bryan, L. A. 2003. Nestedness of beliefs: Examining a prospective elementary teacher's belief system about science teaching and learning. *Journal of Research in Science Teaching* 40 (9): 835–868.

Business Higher Education Forum (BHEF). 2010. Increasing the number of STEM graduates: Insights from the U.S. STEM education and modeling project. *www.bhef.com/sites/bhef.drupalgardens.com/files/report_2010_increasing_the_number_of_stem_grads.pdf*

Business Higher Education Forum (BHEF). 2007. An American imperative: Transforming the recruitment, retention, and renewal of our nation's mathematics and science teaching workforce. *www.bhef.com/sites/bhef.drupalgardens.com/files/report_2007_american_imperative.pdf.*

Carnegie Foundation of New York (CFNY). 2009. *The opportunity equation: Transforming mathematics and science education for citizenship and the global economy*. New York: CFNY.

Casey, J. 1994. TeacherNet: Student teachers travel the information highway. *Journal of Computing in Teacher Education* 11 (1): 8–11.

Cook, C. 1991. *Professional development interview/survey questions*. Oak Brook, IL: North Central Regional Educational Laboratory.

Cook, C., and C. Rasmussen. 1994. *Framework for designing effective professional development: Change-based inquiry process*. Oak Brook, IL: North Central Regional Educational Laboratory.

Corcoran, T. B. 1995a. Transforming professional development for teachers: A guide for state policymakers. Washington, DC: National Governors' Association.

Corcoran, T. B. 1995b. Helping teachers teach well: Transforming professional development. Policy brief no. RB-16. New Brunswick, NJ: Rutgers University, Consortium for Policy Research in Education.

Corcoran, T. B. 2007. Teaching matters: How state and local policymakers can improve the quality of teachers and teaching. Policy brief no. RB-48. New Brunswick, NJ: Rutgers University, Consortium for Policy Research in Education.

Corcoran, T., and M. Goertz. 1995. Instructional capacity and high performance. *Educational Researcher* 24 (9): 27–31.

Corcoran, T. B., P. M. Shields, and A. A. Zucker. 1998. *Evaluation of NSF's Statewide Systemic Initiatives (SSI) program: The SSIs and professional development for teachers*. Menlo Park, CA: SRI International.

Crowe, E. 2010. *Measuring what matters. A stronger accountability model for teacher education.* Washington, DC: The Center for American Progress.

Cuban, L. 1990. Reforming again, again, and again. *Educational Researcher* 19 (1): 3–13.

Cuban, L. 1992. What happens to reforms that last? The case of the junior high school. *American Educational Research Journal* 29 (2): 227–251.

Darling-Hammond, L. 2010. Teacher Education and the American Future. *Journal of Teacher Education* 61 (1-2): 35–47.

Darling-Hammond, L., and M. W. McLaughlin. 1995. Policies that support professional development in an era of reform. *Phi Delta Kappan* 76 (8): 597–604.

Davis, K. 2002. "Change is hard": What science teachers are telling us about reform and teacher learning of innovative practices. *Science Education* 87 (1): 3–30.

DeBoer, G. 1991. *A history of ideas in science education: Implications for practice.* New York City: Teachers College Press.

DeBoer, G. 2000. Scientific literacy: Another look at its historical and contemporary meanings and its relationship to science education reform. *Journal of Research in Science Teaching* 37 (6): 582–601.

Earley, P., and V. Porritt. Forthcoming. Evaluating the impact of professional development: The need for a student-focused approach. *Professional Development in Education.*

Edelson, D. C., D. N. Gordin, and R. D. Pea. 1999. Addressing the challenges of inquiry-based learning through technology and curriculum design. *Journal of the Learning Sciences* 8 (3-4): 391–450.

Elbaz, F. 1983. *Teacher thinking: A study of practical knowledge.* London: Croom Helm.

Eraut, M. 1988. Knowledge creation and knowledge use in professional contexts. *Studies in Higher Education* 10 (2): 117–132.

Eraut, M. 1994. *Developing professional knowledge and competence.* London: Falmer Press.

Feldman, A. 2000. Decision-making in the practical domain: A model of practical conceptual change. *Science Education* 84 (5): 606–623.

Fields, E. T., A. J. Levy, M. K. Tzur, A. Martinez-Gudapakkam, and E. Jablonski. 2012. The science of professional development. *Phi Delta Kappan* 93 (8): 44–46.

Fine, C. 1994. *Breaking out of the egg crates: Redesigning professional roles and relationships.* PhD diss. National-Louis University.

Fullan, M. 1993. *Change forces: Probing the depths of educational reform.* London: Falmer Press.

Fullan, M. G. 2007a. Change the term for teacher learning. *Journal of Staff Development* 28 (3): 35–36.

Fullan, M. G. 2007b. *The new meaning of educational change.* 4th ed. New York City: Teachers College Press.

Fullan, M. G., and M. Miles. 1992. Getting reform right: What works and what doesn't. *Phi Delta Kappan* 73 (10): 745–752.

Fullan, M., and S. Steigelbauer. 1991. *The new meaning of educational change.* 2nd ed. New York: Teachers College Press.

Furtak, E. M. 2012. Linking a learning progression for natural selection to teachers' enactment of formative assessment. *Journal of Research in Science Teaching* 49 (9): 1181–1210.

Gay, G. 2002. Preparing for culturally responsive teaching. *Journal of Teacher Education* 53 (2): 106–116.

Goal 2000: Educate America Act, Pub. L. No. 103-227, 108 Stat. 125 (1994).

Grimberg, B. I., and E. Gummer. 2013. Teaching science from cultural points of intersection. *Journal of Research in Science Teaching* 50 (1): 12–32.

Guskey, T. R. 1982. The effects of change in instructional effectiveness upon the relationship of teacher expectations and student achievement. *Journal of Educational Research* 75 (6): 345–349.

Guskey, T. R. 1984. The influence of change in instructional effectiveness upon the affective characteristics of teachers. *American Educational Research Journal* 21 (2): 245–259.

Guskey, T. R. 1985. Staff development and teacher change. *Educational Leadership* 42 (7): 57– 60.

Guskey, T. R. 1986. Staff development and the process of teacher change. *Educational Researcher* 15 (5): 5–12.

Guskey, T. R. 1989. Attitude and perceptual change in teachers. *International Journal of Educational Research* 13 (4): 439–453

Guskey, T. R. 1994. Results-oriented professional development: In search of an optimal mix of effective practices. *Journal of Staff Development* 15 (4): 42–50.

Guskey, T. R. 2002. Does it make a difference? Evaluating professional development. *Educational Leadership* 59 (6): 45–51.

Guskey, T. R., and K. S. Yoon. 2009. What works in professional development? *Phi Delta Kappan* 90 (7): 495–500.

Hall, G. E., and S. Loucks. 1978. Teacher concerns as a basis for facilitating and personalizing staff development. *Teachers College Record*, 80: 36–53.

Hall, G. E., and S. Loucks. 1979. *Implementing innovations in schools: A concerns-based approach.* Austin, TX: Research and Development Center for Teacher Education, University of Texas.

Herold, B. "Benefits of online, face-to-face professional development similar, study finds," *Digital education* (blog), *Education Week*, June 19, 2013, *blogs.edweek.org/edweek/DigitalEducation/2013/06/no_difference_between_online_a.html.*

Huberman, M. 1985. What knowledge is of most worth to teachers? A knowledge-use perspective. *Teaching and Teacher Education* 1 (3): 251–262.

Huberman, M. 1995. Professional careers and professional development: Some intersections. In *Professional development in education: New paradigms and practices*, ed. T. R. Guskey and M. Huberman, 193–224. New York: Teachers College Press.

Huberman, M., and M. M. Grounauer. 1993. *The lives of teachers*. New York: Teachers College Press.

Ingersoll, R. M. 2011. Do we produce enough mathematics and science teachers? *Phi Delta Kappan* 92 (6): 37–41.

Ingersoll, R., and L. Merrill. 2010. Who's teaching our children? *Educational Leadership* 67 (8): 14–20.

Johnson, C. C., J. B. Kahle, and J. D. Fargo. 2007. A study of the effect of sustained, whole school professional development on student achievement in science. *Journal of Research in Science Teaching* 44 (6): 775–786.

Joyce, B., and M. Weil. 1980. *Models of teaching*. Englewood Cliffs. NJ: Prentice Hall

Kennedy, M. 1998. The relevance of content in inservice teacher education. Paper presented at the annual meeting of the American Educational Research Association, San Diego, California.

Killion, J., and C. Harrison. 1997. The multiple roles of staff developers. *Journal of Staff Development* 18 (3): 33–44.

Knowles, M. 1980. *The modern practice of adult education.* Chicago, IL: Association/Follett Press.

Kober, N., and A. Usher. 2012. *Public education primer: Basic (and sometimes surprising facts) about the U.S. education system, 2012 revised edition.* Washington, DC: Center on Education Policy.

Lee, O., J. E. Hart, P. Cuevas, and C. Enders. 2004. Professional development in inquiry-based science for elementary teachers of diverse student groups. *Journal of Research in Science Teaching* 41 (10): 1021–1043.

Lieberman, A. 1995. Practices that support teacher development: Transforming conceptions of professional learning. *Phi Delta Kappan* 76 (8): 591–596.

Lieberman, A., and M. W. McLaughlin. 1992. Networks for educational change: Powerful and problematic. *Phi Delta Kappan* 73 (9): 673–677.

Little, J. W. 1987. Teachers as colleagues. In *Educators' handbook: A research perspective,* ed. V. Richardson-Koehler, 491–518. New York: Longman Press.

Lockwood, A. T. 1995. Constructing communities of cooperation. *New Leaders for Urban Schools* 1 (1): 3–7.

Lord, B. 1994. Teachers' professional development: Critical colleagueship and the role of professional communities. In *The future of education: Perspectives on national standards in education,* ed. N. Cobb, 175–204. New York: College Entrance Examination Board.

Loucks-Horsley, S. 1995. What the professional developer/designer does. Paper presented at the Education Development Center's Conference for professional development teams for the 25 Statewide Systemic Initiatives, Baltimore, MD.

Loucks-Horsley, S., and R. W. Bybee. 1998. Implementing the national science education standards: How will we know when we're there? *Science Teacher* 65 (6): 22–25.

Loucks-Horsley, S., P. Hewson, N. Love, and K. Stiles. 1998. *Designing professional development for teachers of science and mathematics.* Thousand Oaks, CA: Corwin Press.

Loucks-Horsley, S. P., N. B. Love, K. E. Stiles, S. Mundry, and P. W. Hewson. 2003. *Designing Professional development for teachers of science and mathematics.* 2nd ed. Thousand Oaks, CA. Corwin Press.

Loucks-Horsley, S., K. E. Stiles, S. Mundry, N. Love, and P. Hewson. 2009. *Designing professional development for teachers of science and mathematics.* 3rd ed. Thousand Oaks, CA: Corwin Press.

Loucks-Horsley, S., K. Stiles, and P. Hewson. 1996. *Principles of effective professional development for mathematics and science education: A synthesis of standards.* Madison, WI: National Institute for Science Education, University of Wisconsin-Madison (ERIC Document Reproduction no. ED 409 201)

Loveless, T. 2011. The 2010 Brown Center report on American education: How well are American students learning? Washington, DC: Brookings Institution. *www.brookings.edu/~/media/research/files/reports/2011/2/07%20education%20loveless/0207_education_loveless.pdf*

Marsh, D. D., and K. LeFever. 1997. Educational leadership in a policy context. What happens when student performance standards are clear? Paper presented at the annual meeting of the American Educational Research Association, Chicago, IL.

Marx, R. W., P. C. Brumenfeld, J. S. Krajcik, B. Fishman, E. Soloway, R. Geier, and R. T. Tal. 2004. Inquiry-based science in the middle grades: Assessment of learning in urban systemic reform. *Journal of Research in Science Teaching* 41 (10): 1063–1080.

Maskiewicz, A. C., and V. A. Winters. 2012. Understanding the co-construction of inquiry-based practices: A case study of a responsive teaching environment. *Journal of Research in Science Teaching* 49 (4): 429–464.

McLaughlin, M. W. 1991. Enabling professional development: What have we learned? In *Staff development for education in the '90s,* ed. A. Lieberman and L. Miller, 61–82. New York: Teachers College Press.

McLaughlin, M. W., and D. D. Marsh. 1978. Staff development and school change. *Teachers College Record* 80 (1): 69–93.

Metz, K. 2008. Elementary school teachers as "targets and agents of change": Teachers' learning in interaction with reform science curriculum. *Science Education* 93 (5): 915–954.

Murphy, C. 1992. Study groups foster schoolwide learning. *Educational Leadership* 50 (3): 71–74.

Mutch-Jones, K., G. Puttick, and D. Minner. 2012. Lesson study for accessible science: Building expertise to improve practice in inclusive science classrooms. *Journal of Research in Science Teaching* 49 (8): 1012–1034.

National Center for Education Statistics (NCES). 2001. *Highlights from the Third International Mathematics and Science Study–Repeat (TIMSS–R).* NCES 2001–027. Washington, DC: National Center for Education Statistics, U.S. Department of Education.

National Center for Education Statistics (NCES). 2009. *The nation's report card: National Assessment of Educational Progress (NAEP) at grade 8, science 2009.* Washington, DC: NCES, U.S. Department of Education.

National Center for Educational Statistics (NCES). 2011. *The nation's report card: National Assessment of Educational Progress (NAEP) at grade 8, science 2011.* Washington, DC: NCES, Institute of Education Sciences, U.S. Department of Education.

National Commission on Excellence in Education (NCEE). 1983. *A nation at risk: The imperative for educational reform.* Washington, DC: U.S. Department of Education.

National Research Council (NRC). 1996. *National science education standards*. Washington, DC: National Academies Press.

National Science Board (NSB). 2000. *Science and engineering indicators (SEI)*, vol 1. Washington, DC: National Science Foundation.

Newmann, F. N. 1994. School-wide professional community. *Issues in Restructuring Schools* 6: 1–2.

NGSS Lead States. 2013. *Next Generation Science Standards: For states, by states*. Washington, DC: National Academies Press. *www.nextgenscience.org/next-generation-science-standards.*

Nias, J. 1989. Teaching and the self. In *Perspectives on teachers' professional development*, ed. M. L. Holly, and C. S. McLoughlin, 155–171. London: Falmer Press.

O'Day, J., and M. Smith. 1993. Systemic reform and educational opportunity. In *Designing coherent education policy: Improving the system*, ed. S. H. Fuhrman, 250–312. San Francisco: Jossey-Bass.

Parke, H. M., and C. R. Coble. 1997. Teachers designing curriculum as professional development: A model for transformational science teaching. *Journal of Research in Science Teaching* 34 (8): 773–789.

Pennell, J. R., and W. A. Firestone. 1996. Changing classroom practices through teacher networks: Matching program features with teacher characteristics and circumstances. *Teachers College Record* 98 (1): 46–76.

Porter, A., B. Birman, M. S. Garet, L. M. Desimone, and K. S. Yoon. 2004. *Effective professional development in mathematics and science: Lessons from evaluation of the Eisenhower Program.* Washington, DC: American Institutes of Research.

Porter, A. C., M. S. Garet, L. Desimone, and B. Birman. 2004. *Designing professional development that works.* Washington, DC: U. S. Department of Education.

President's Council of Advisors on Science and Technology (PCAST). 2010. Prepare and inspire: K–12 education in science, technology, engineering, and math (STEM) education for America's future. *www.whitehouse.gov/sites/default/files/microsites/ostp/pcast-stem-ed-final.pdf.*

President's Council of Advisors on Science and Technology (PCAST). 2012. Report to the President. Engage to excel: Producing one million additional college graduates with degrees in science, technology, engineering, and mathematics. *www.whitehouse.gov/sites/default/files/microsites/ostp/pcast-engage-to-excel-final_2-25-12.pdf.*

Rosebery, A. S., B. Warren, and F. R. Conant. 1989. *Making sense of science in language minority classrooms.* BBN tech. rep. no. 7306. Cambridge, MA: Bolt, Beranek, and Newman.

Rosebery, A. S., B. Warren, and F. R. Conant. 1992. *Appropriating scientific discourse: Findings from language minority classrooms.* NCRCDSLL research report 3. Berkeley, CA: National Center for Research on Cultural Diversity and Second Language Learning, UC Berkeley.

Roth, K. J., H. E. Garnier, C. Chen, M. Lemmens, K. Schwille, and N. I. Wickler. 2011. Videobased lesson analysis: Effective science PD for teacher and student learning. *Journal of Research in Science Teaching* 48 (2): 117–148.

Schneider, R. M., J. Krajcik, and P. Blumenfeld. 2005. Enacting reform-based science materials: The range of teacher enactments in reform classrooms. *Journal of Research in Science Teaching* 42 (3): 283–312.

Schwartz, M. S., P. M. Sadler, G. Sonnert, and R. H. Tai. 2008. Depth versus breadth: How content coverage in high school science courses relates to later success in college science coursework. *Science Education* 93 (5): 798–826.

Senge, P. M. 1990. *The fifth discipline: The art and practice of the learning organization.* New York: Doubleday.

Shulman, L. S. 1987. Knowledge and teaching: Foundations of the new reform. *Harvard Educational Review* 57 (1): 1–22.

Smith, M. S., and J. O'Day. 1991. Putting the pieces together: Systemic school reform. Policy brief RB-06-4/91. New Brunswick, NJ: Consortium for Policy Research in Education.

Smylie, M. A. 1989. Teachers' views of the effectiveness of sources of learning to teach. *Elementary School Journal* 89 (5): 543–558.

Sparks, D., and S. Loucks-Horsley. 1989. Five models of staff development for teachers. *Journal of Staff Development* 10 (4): 40–57.

Spillane, J. P. 2000. District leaders' perceptions of teacher learning. CPRE occasional paper series OP-05. Philadelpha, PA: Consortium for Policy Research in Education.

Spillane, J., and K. A. Callahan. 2000. Implementing state standards for science education: What district policymakers make of the hoopla. *Journal of Research in Science Teaching* 37 (5): 401–425.

Supovitz, J. A., and H. M. Turner. 2000. The effects of professional development on science teaching practices and classroom culture. *Journal of Research in Science Teaching* 37 (9): 963–980.

Trumbull, E., and M. Pacheco. 2005. *Leading with diversity: Cultural competencies for teacher preparation and professional development.* Providence, RI: The Education Alliance at Brown University.

U. S. Department of Education (USED). 2002. *Meeting the highly qualified teacher challenge.* Washington, DC: U. S. Department of Education, Office of Postsecondary Education, Office of Policy Planning and Innovation. *www.ed.gov/offices/OPE/News/teacherprep/AnnualReport.pdf.*

U. S. Department of Education (USED). 2010. A blueprint for reform: Reauthorization of the Elementary and Secondary Education act. *www2.ed.gov/policy/elsec/leg/blueprint/blueprint.pdf.*

Wei, R. C., L. Darling-Hammond, and F. Adamson. 2009. *Professional development in the United States: Trends and challenges.* Dallas, TX: Learning Forward/National Staff Development Council.

Wei, R. C., L. Darling-Hammond, A. Andree, N. Richardson, and S. Orphanos. 2009. *Professional learning in the learning profession: A status report on teacher development in the United States and abroad.* Dallas, TX. National Staff Development Council.

Yin, R. 2006. *Cross-site evaluation of the Urban Systemic Program. The final annual report: Baseline outcome analysis.* Washington, DC: COSMOS Corporation.

Yoon, K. S., T. Duncan, S. Lee, B. Scarloss, and K. Shapley. 2007. Reviewing the evidence on how teacher professional development affects student achievement. Issues and answers report, REL 2007—no. 033. Washington, DC: U.S. Department of Education, Institute of Education Sciences, National Center for Education Evaluation and Regional Assistance, Regional Educational Laboratory Southwest.

Chapter 3

The Importance of Viable Models in the Construction of Professional Development

Joseph Krajcik

We live in an exciting time in science and science education. Over the last 10 years, many amazing new scientific breakthroughs have occurred that impact our daily lives: genomics, nanoscience, and the use of digital technologies for communications, to name just a few. These breakthroughs give us more control over serious illnesses and allow us to communicate globally through pictures, voice, and text in real time using handheld technologies and to travel the globe within a day. While amazing and useful, these scientific breakthroughs give rise to many technical, ethical, and moral problems such as global warming, pollution of waterways and the air, decrease and loss of species, and a dwindling supply of energy and other resources. Hence, the children of today will grow up in a world in which they will need to apply scientific concepts, communicate ideas, make sound decisions based on evidence, and collaborate with others to solve these problems and prevent them from escalating.

In the past 15 years, learning and cognitive scientists have made tremendous advances in our understanding about how students learn science and how science should be taught to help prepare students for the rapidly changing world. These ideas have been well documented in several publications by the National Research Council (NRC), such as *How Students Learn: History, Mathematics, and Science in the Classroom* (Bransford and Donovan 2005), *Knowing What Students Know* (Pelligrino, Chudowsky, and Glaser 2001), *Taking Science to School* (Duschl, Schweingruber, and Shouse 2007), *Ready, Set, Science!* (Michaels, Shouse, and Schweingruber 2008), *America's Lab Report* (Singer, Hilton, and Schweingruber 2005), *Successful K–12 STEM Education: Identifying Effective Approaches in Science Technology, Engineering, and Mathematics* (NRC 2011) and *A Framework for K–12 Science Education* (*Framework*; NRC 2012). The findings reported in these publications clearly show that to be productive 21st-century global citizens, learners need to develop integrated understanding of big ideas of science by applying and using big ideas to explain phenomena and solve problems important to them. By "integrated understanding" I mean that ideas are linked together in a weblike fashion that allows learners to access information for problem solving and decision making (Fortus and Krajcik 2011).

The *Framework* provides a coherent picture of the major scientific and engineering ideas and practices that all learners need to understand in order to live productive lives as citizens in this century and, if desired, to pursue further study of science and engineering. The *Framework* makes use of four key ideas: (1) a limited number of big ideas of science, (2) an ongoing developmental process, (3) the integration or coupling of core ideas and scientific practices, and (4) crosscutting elements.

The *Framework* laid the foundation for the *Next Generation Science Standards* (*NGSS*; NGSS Lead States 2013), in which we find the blending of core ideas with scientific practices and crosscutting concepts that is central to the *NGSS*. With these breakthroughs in science education, the implications are clear for what inservice and preservice teachers must do to be prepared to teach new scientific ideas using sound pedagogical methods to support students. Taken collectively, the country needs to develop and institute a nationwide approach aimed at preparing top-notch K–12 science teachers before they enter the teaching profession and an equally effective program for providing high-quality professional development to practicing teachers, regardless of the route taken.

How can we use the *Framework* and the *NGSS* to inform teaching and learning and its concomitant professional development? These publications will likely serve as the core that education stakeholders at all levels will rally around to establish such an educational infrastructure. The documents clearly demonstrate that what we teach needs to change because of what we know. Rather than focusing on multiple ideas, the *Framework* recommends that teachers help students develop understanding of the core ideas of science because these will help learners form a foundation for lifelong learning.

As an example of this process and its implications, one of the core ideas in physical science in the *Framework* is Energy, and one of the crosscutting concepts is Energy and Matter. All teachers from kindergarten through high school will need to present a coherent vision of energy. Previously, the concept of energy was relegated to physical science courses. As such, students had a hard time seeing the similarities of the energy discussed in chemistry and biology with the transformation of kinetic energy to gravitational energy in physics class. Within our own teaching, we create a schism, whereas, energy is in reality a crosscutting concept essential to all of the disciplines. As the *Framework* stresses, the idea of energy is essential in examining the systems of life science, Earth and space science, chemical systems, and engineering contexts. The idea of energy needs to be taught not just in physics or physical science, but across the grade levels and integrated into all of the science, technology, engineering, and mathematics (STEM) subjects.

Energy is difficult to define, yet we can track energy as it transfers across various systems. Since many teachers have never been taught to teach energy in this powerful way, this approach will need to be embedded into professional development at all levels. Elementary teachers will need to be able to introduce the idea of energy in ways that middle and high school science teachers can build on to further help students develop deeper and more powerful ideas of energy that can be used to explain phenomena and solve problems.

What model of professional development can we use that will help teachers at the elementary level develop an understanding of energy while also providing the tools to support elementary students beginning to form an understanding of energy? How do we support middle school teachers in developing a deep and integrated understanding of energy across the grades and the various disciplines and to link the ideas together within and across the grade levels?

With respect to how we teach science, many classrooms in the United States still resemble classrooms of the early1900s, with outdated equipment and pedagogical strategies that fail to promote learning for most students. John Dewey bemoaned, in 1910, that education focuses too much on facts and not on how knowledge is generated. We heard a similar cry from Schwab in the 1960s and again from Bruce Alberts in 2009. Although we have seen some changes, learning science is still too much like learning a language and not enough about explaining how the world works. Teaching in which ideas build upon each other is in many ways a foreign idea to science teachers. Although it is a hallmark of teaching and learning because it ties to the importance of connecting to prior knowledge, linking conceptual ideas across time is a challenging pedagogical practice that is seldom observed in the teaching of science. Yet, it can have a powerful influence on student learning (Roseman, Linn, and Koppal 2008).

Today we know much about how to engage students in constructing, revising, and communicating models and in building and communicating explanations from evidence. Students need to engage in model construction and revision as well as in the building and communication of explanations based on new evidence in order to explain phenomena (NRC 2011; Krajcik and Merritt 2012; McNeill and Krajcik 2011). For the science education community, beyond providing a newer student-centered approach to professional development, another major challenge lies in our nation's large urban cities and rural areas in which classrooms are increasingly filled with underrepresented populations from a variety of cultures (e.g., Hispanics, African Americans, Asians). As a nation, we face the tremendous challenge of how to provide quality science education to diverse learners whose culture and ways of knowing may vary significantly from those of their teachers, necessitating professional development that includes ways to address the cultural and linguistic issues inherent to such classrooms (Moje et al. 2001). The question then becomes how do we support teachers in learning and enacting these important scientific practices?

The Problem

Although the *Framework* and the *NGSS* are critical steps in upgrading science education in the United States, their impact will be limited by the degree to which K–12 teachers implement the *NGSS* with fidelity. Unfortunately, past experience with adapting standards suggests that implementing the *NGSS* as intended by their developers may be compromised because of the ways local school districts interpret the standards (Spillane and Callahan 2002). For example, we can expect the terms "learning progressions" and "scientific practices" may take on a wide range of meanings as they are introduced to teachers and school personnel. Many of these interpretations will diverge from the *Framework* committee's intentions. Teachers also have a tendency to judge their activities as aligned to standards even when "diverging widely" from that which was intended (Penuel et al. 2009, p. 28). For instance, many teachers felt they were doing inquiry in the classroom when students did a hands-on activity, but hands-on does not necessarily equate to doing inquiry. Moreover, the *Framework* and the *NGSS* will not be sufficient resources for helping teachers learn about and enact the standards. Teachers will need professional development in how to interpret and implement the next generation of standards in their teaching, and until new teaching materials are developed, teachers will need to know how to blend the ideas in the *NGSS* with their current learning materials.

How do we prepare teachers of science, particularly those who graduated in the past, with new scientific ideas and new understandings of how to teach? Unfortunately, many science teachers in our schools have not continued in their professional growth. There are many reasons for teachers not taking part in professional development in our country, including lack of national and state policies that provide both financial backing and time for this commitment. Yet, taking part in professional growth opportunities to learn new science ideas and new methods on how to teach children is critical. The infrastructure for professional development in science education has changed significantly since the first generation of science standards. Given the limited time available for face-to-face professional development and the increasing budget constraints for carrying out professional development, what viable models can be used to support teachers in enacting *and understanding* essential features of the next generation of science standards? What innovative curriculum and new teaching ideas have emerged in the field? How do we support teachers in developing the pedagogical content knowledge they need to help learners? As a nation, we need to build cost-effective, scalable, accessible professional development models that can support teachers in understanding innovations such as the vision painted in the *Framework* and the *NGSS*.

How important is professional development? Forty years ago, when I was in my late teens and early twenties, I was a good car mechanic. I felt competent to open the hood or crawl under most cars to fix them. I even put a new clutch (not a simple procedure) into my Volkswagen bug. But today, I can't even find the batteries in most cars. Although cars today look superficially like they did in 1970 and still mostly run on internal combustion engines, the internal workings of most new cars are based on updated computer technologies. I never kept up my education as a car mechanic, and with changes in the design and running of cars, my knowledge is old and outdated. Although I could probably still change the spark plugs in a 1968 Volkswagen bug, that knowledge is no longer useful for today's cars. Today I would have no clue about what to do if my car stalled except to make a phone call. The same thing is true of teaching. Just as you would not want me working on your 2012 Volkswagon because I no longer have that knowledge or skill, many teachers do not have the knowledge and teaching skills for teaching in today's classrooms. They did not keep up with their professional development. What we teach and how we teach has changed. While classrooms superficially look the same, what we do in them needs to be very different.

The NRC in *Taking Science to School: Learning and Teaching Science in Grades K–8* (Duschl, Schweingruber, and Shouse 2007) argues that well-designed professional development opportunities for teachers can produce the desired changes in classroom practices and contribute to improvements in student learning. But to do so we need viable models of professional development that we know work. Kubitskey and Fishman (2007) support the statement made by the NRC and propose a model of how professional development can influence student outcomes. They believe professional development activities can influence teacher buy-in to an innovation and to the teacher knowledge and confidence needed for the innovation to occur. These components are critical to the practices that teachers use in the classroom and that, in turn, will influence student learning. What is also critical is that we learn more about what type of professional development activities are most promising.

Unfortunately, although we have gained knowledge of what can work, the field lacks knowledge in how to scale and support the use of these new ideas. Typically, professional development institutes do not allow time to provide teachers with the background necessary to teach new ideas. Often ideas are presented superficially and the rationale behind the idea and the importance of using the ideas with fidelity is not stressed. Too often, professional development focuses on presenting the innovations to teachers without engaging them in the process. As such, many science teachers often adapt their use of innovation based on their prior knowledge of teaching and learning, which causes the new methods to resemble traditional classroom practices.

Building Professional Development Models From What Is Known

Professional development that supports teachers' learning has been shown to be a key factor in improving the quality of schools (e.g., Borko and Putnam 1995) and student learning (Desimone et al. 2005; Heller et al. 2012). Previous research indicates that professional development for science teachers needs to have several key features, including clearly specified learning goals that focus on instruction and student outcomes and highly interactive sessions that engage teachers in a community that supports their learning (Darling-Hammond 1997). Professional development must also provide opportunities for collective meaning making and focus on authentic problems from the teachers' perspective. Moreover, we know professional development needs to be sustained over a long period. Such professional development can lead to desired changes in teacher knowledge and practice (Penuel et al. 2007).

Lee and Krajcik (2012) suggest a viable model to develop teacher knowledge and teaching skills through a combination of effective professional development and educative materials embedded in the curriculum (Davis and Krajcik 2005; Remillard 2005). Educative materials build supports into teaching materials that allow teachers to enact the innovation as intended.

Yet, what goes into these professional development models? What features of professional development should be focused on? Moreover, we need evidence to support these models.

A Model Supported by Research

Here I discuss a viable model of professional development that stems from the work of Joan Heller and colleagues (Heller et al. 2012). Heller and colleagues used a randomized experimental design, implemented in six states with over 270 elementary teachers and 7,000 students in order to compare three related but systematically varied teacher professional development interventions (*Teaching Cases, Looking at Student Work,* and *Metacognitive Analysis*) along with a no-treatment control group. The three interventions contained similar science content components but differed in the ways they incorporated analysis of learner thinking and of teaching. Another critical aspect of their design involved facilitators not involved in the design of the interventions to deliver the professional development sessions. This design made it possible to measure effects of the unique feature of each intervention on teacher and student outcomes. The findings indicate that each intervention improved teachers' and students' scores on selected-response science tests significantly and substantially beyond those of control students, and the effects lasted until a year later. Student achievement also improved significantly for English language learners in both the study year and follow-up, with the intervention effects not differing based on sex, race, or ethnicity. However, the research team did see important differences resulting from the various interventions. Only the *Teaching Cases* and *Looking at Student Work* interventions improved the accuracy and completeness of students' written justifications of test answers on follow-up assessments, and only *Teaching Cases* had sustained effects on teachers' written justifications. Although the content component that was common across the three interventions showed powerful effects on teachers' and students' ability to select correct test answers, the ability to explain why answers were correct only improved when the professional development incorporated analysis of student conceptual understandings and implications for instruction.

These findings are important for several reasons: First, they show that professional development that integrates content learning with analysis of student learning and implications for instruction can impact student learning. Second, the study demonstrated that high-quality professional development of moderate duration can be delivered by facilitators not involved in the development of the interventions and can have considerable and lasting impact on the teaching and learning of elementary science. In addition to the impact, points one and two are important because they illustrate that other professional development models need to incorporate and test these components. Third, the effects of the interventions were stronger for teachers' students in the follow-up year, suggesting that teachers need to have several iterations before students in their classrooms experience the full impact of the professional development. Often in measuring professional development, we take one-shot approaches that don't allow teachers opportunities to develop their understanding further, and we do not measure impact across years. Fourth, only a few studies have shown a causal link between the professional development intervention and student outcomes. The study by Heller and colleagues provides an excellent example for others in the field to replicate at different grade levels and for different content. As such, their model provides an excellent example of an effectiveness study in the cycle of development and research.

A Professional Development Model to Support Enactment of Innovative Curriculum

Using learning goals–driven design (Krajcik, McNeill, and Resier 2008), we developed a middle school science curriculum for grades 6–8 with curriculum coherence as a central design principle (Shwartz et al. 2008). *Investigating and Questioning Our World Through Science and Technology* (IQWST; Krajcik et al. 2011) is a project-based curriculum comprising biology, chemistry, physics, and Earth science units that focuses on building big ideas across time, using scientific practices and engaging students in explaining phenomena. Because IQWST stresses the development of big ideas blended with scientific practices across time, it matches closely the ideas in the *Framework*. This brings challenges not only in introducing teachers to a new curriculum but also to engaging teachers in learning core ideas blended with scientific practices. How do you design a weeklong professional development institute to support teachers in enacting a yearlong, project-based curriculum in which ideas build on each other?

Our model focused on engaging teachers in pedagogy and practices common to all units during a one-week summer institute in order to provide teachers with generalizable knowledge for teaching a full year of IQWST (Krajcik et al. 2008). We learned early on that it is important to collaborate with teachers in designing professional development experiences and to engage teachers in the doing of science during these experiences (Krajcik et al. 1994). The summer institute was followed by a two-day, unit-specific professional development preceding the teaching of each new IQWST unit to reinforce the generalizable ideas in the new context. A key aspect of our work involved teachers in experiencing the materials through model teaching and reflection.

The generalizable knowledge we intended teachers to come away with included

- contextualizing learning using a driving question and a driving question board to frame each unit, and providing a series of investigable questions that motivate students with a need to know;
- scaffolding specific scientific practices such as creating and testing scientific models and constructing scientific explanations as important approaches to classroom inquiry;
- reinforcing classroom learning by providing students with age-appropriate, expository text written to support a range of learners as they read about science in and out of the classroom and engage in multiple ways of expressing their understanding;

- focusing on helping students develop a deep understanding of each unit's learning goals through a coherent instructional sequence and showing teachers how to link ideas within units using the driving questions and driving question boards; and
- fostering a collaborative classroom culture by focusing on specific types of interactive classroom discussions.

Assessment of teacher artifacts created during the workshop indicated that teachers recognized and appropriately described key features of IQWST and that they began to realize the challenges of implementing the materials in their own classrooms. The artifacts also revealed that the teachers constructed evidence-based scientific models by experiencing phenomena and engaging in lessons on how to construct evidenced-based models. Analysis of workshop records further indicated that teachers engaged in alternate classroom discourse patterns described by the facilitators and used in IQWST materials. Teachers were also able to identify key IQWST features, explain their importance, and describe their associated challenges in enactment. During the workshop, teachers discussed challenges to implementing IQWST, such as facilitating discussions, differentiating instruction, and engaging students to develop evidence-based classroom models over a series of lessons.

The findings from this study suggest that, given the limited time teachers are able to dedicate to professional development experiences, focusing on generalizable pedagogical components of reform-based materials—those that are not specific to units but that apply across units and could be used in other situations as they are tied to what is known about student learning—may be an important way to shape teacher practice. Our work also illustrates that engaging teachers through modeling teaching and engaging with materials is essential to learning. However, this research is at the design, develop, and test phase of the cycle. Although we have some evidence that focusing on generalizable principles can support teachers, we need to take this research to the next phase of the cycle.

Uses of Technology

Given the current state of limited time and resources juxtaposed against the number of teachers that need professional development, technological advances in using synchronous and nonsynchronous communication become invaluable in improving the knowledge and skills of teaching. One of the greatest advantages of using a technological delivery method is that one can impact a large number of teachers at any given time. With the cost of tablet computers lowering to affordable price levels and connectivity speeds increasing, high-quality professional development could be delivered on an as-needed basis through interactive online sources. Under this condition, educators can watch videos of experienced teachers enacting lessons, see how to perform investigations, watch how to support student-to-student dialogue, and learn about challenges students face in learning the ideas. Although online interactive materials are expensive to develop, many institutions of higher education, educational nonprofit organizations, and educational technological-based firms are actively meeting the challenge. It is only through multiple avenues for meeting the demands for effective professional development that we will see results in subsequent classroom practices and student achievement.

A New Professional Development Model Under Study

To support teachers in enacting the *NGSS*, my colleagues and I are developing, testing, and revising a model for professional development for the *Framework* and the *NGSS* on the basis of what we know about effective professional development. The iterative professional development model is intended to support teachers in

- developing understanding of what is meant by core ideas, scientific practices, and crosscutting concepts.
- developing understanding of how to blend core ideas with scientific practices and crosscutting concepts to develop learning performances or learning goals. This is the same process that will be used to create the next generation of standards.
- modifying existing instructional materials by identifying core ideas, scientific practices, and crosscutting concepts.
- developing learning tasks and assessment measures that will meet learning performances or learning goals.

Although we have implemented this model once with some degree of effectiveness, on the basis of responses from teachers, we know we are only at the beginning stages of our logic model for research and development.

Concluding Comment

With the release of the *Framework* and the *NGGS*, we have the opportunity to improve the teaching and learning of science and to help move our nation forward by providing all students with the depth of understanding of big ideas, scientific and engineering practices, and crosscutting elements needed to be productive citizens and leaders in the 21st century. However, extensive professional development based on viable models supported by careful research is needed in order for teachers to implement the vision outlined in the *Framework* and the *NGSS* with fidelity in science classrooms.

Several features of such professional development discussed in this chapter include

- engaging teachers in analyzing student work,
- engaging teachers in learning generalizable features of new curriculum materials, and
- building teachers' understanding over time.

References

Borko, H., and R. Putnam. 1995. Expanding a teachers' knowledge base: A cognitive psychological perspective on professional development. In *Professional development in education: New paradigms and practices*, ed. T. Guskey and M. Huberman, 35–66. New York: Teachers College Press.

Bransford, J. D., and M. S. Donovan. 2005. Scientific inquiry and how people learn. In *How students learn: History, mathematics, and science in the classroom*, ed. M. S. Donovan and J. D. Bransford, 397–416. Washington, DC: National Academies Press.

Darling-Hammond, L. 1997. *The right to learn: A blueprint for creating schools that work*. San Francisco: Jossey-Bass.

Davis, E. A., and J. Krajcik. 2005. Designing educative curriculum materials to promote teacher learning. *Educational Researcher* 34 (3): 3–14.

Desimone, L. M., T. M. Smith, S. Hayes, and D. Frisvold. 2005. Beyond accountability and average math scores: Relating multiple state education policy attributes to changes in student achievement in procedural knowledge, conceptual understanding and problem solving in mathematics. *Educational Measurement: Issues and Practice* 24 (4): 5–18.

Duschl, R. A., H. A. Schweingruber, and A. W. Shouse, eds. 2007. *Taking science to school: Learning and teaching science in grades K–8*. Washington, DC: National Academies Press.

Fortus, D., and J. Krajcik. 2011. Curriculum coherence and learning progressions. In *The international handbook of research in science education*. 2nd ed, ed. B. J. Fraser, K. G. Tobin, and C. J. McRobbie. Dordrecht, the Netherlands: Springer.

Garet, M. S., A. C. Porter, L. Desimone, B. F. Birman, and K. S. Yoon. 2001. What makes professional development effective? Results from a national sample of teachers. *American Research Journal* 38 (4): 915–945.

Heller, J. I., K. R. Daehler, N. Wong, M. Shinohara, and L. W. Miratrix. 2012. Differential effects of three professional development models on teacher knowledge and student achievement in elementary science. *Journal of Research in Science Teaching* 49 (3): 333–362.

Krajcik, J. S., P. Blumenfeld, R. W. Marx, and E. Soloway. 1994. A collaborative model for helping teachers learn project-based instruction. *Elementary School Journal* 94 (5): 539–551.

Krajcik, J. S., J. A. Fogleman, L. Sutherland, and L. Finn. 2008. Professional development that supports reform: Helping teachers understand and use reform-rich materials. Paper presented at the annual meeting of the American Educational Research Association, New York.

Krajcik, J., K. L. McNeill, and B. Reiser. 2008. Learning-goals–driven design model: Developing curriculum materials that align with national standards and incorporate project-based pedagogy. *Science Education* 92 (1): 1–32.

Krajcik, J., and J. Merritt. 2012. Engaging students in scientific practices: What does constructing and revising models look like in the science classroom? *Science Teacher* 79 (3): 10–13.

Krajcik, J., B. Reiser, L. Sutherland, and D. Fortus. 2011. *IQWST: Investigating and questioning our world through science and technology (middle school science curriculum materials)*. Norwalk, CT: Sangari Global Education/Active Science.

Kubitskey, B., and B. J. Fishman. 2007. A design for using long-term face-to-face workshops to support systemic reform. Paper presented at the American Educational Research Association, Chicago.

Lee, O., and J. Krajcik. 2012. Large-scale interventions in science education for diverse student groups in varied educational settings. *Journal of Research in Science Teaching* 49 (3): 271–280.

Marx, R. W., P. C. Blumenfeld, J. S. Krajcik, and E. Soloway. 1998. New technologies for teacher professional development. *Teaching and Teacher Education* 14 (1): 33–52.

McNeill, K. L., and J. Krajcik. 2011. *Supporting grade 5–8 students in constructing explanations in science: The claim, evidence and reasoning framework for talk and writing*. New York: Pearson, Allyn & Bacon.

Michaels, S., A. W. Shouse, and H. A. Schweingruber, eds. 2008. *Ready, set, science!: Putting research to work in K–8 science classrooms*. Washington, DC: National Academies Press.

Moje, E., T. Collazo, R. Carillo, and R. W. Marx. 2001. "Maestro, what is 'quality'?": Examining competing discourses in project-based science. *Journal of Research in Science Teaching* 38 (4): 469–495.

National Research Council (NRC). 1996. *National science education standards*. Washington, DC: National Academies Press.

National Research Council (NRC). 2011. *Successful K–12 STEM education: Identifying effective approaches in science, technology, engineering, and mathematics*. Washington, DC: National Academies Press.

National Research Council (NRC). 2012. *A framework for K–12 science education: Practices, crosscutting concepts, and core ideas*. Washington, DC: National Academies Press.

National Science Foundation (NSF). 2012. NSF-RAPID: Model for implementing the next generation of science standards. *www.nsf.gov/awardsearch/showAward.do?AwardNumber=1153280*.

NGSS Lead States. 2013. *Next Generation Science Standards: For states, by states*. Washington, DC: National Academies Press. *www.nextgenscience.org/next-generation-science-standards*.

Pelligrino, J. W., N. Chudowsky, and R. Glaser, eds. 2001. *Knowing what students know: The science and design of educational assessment.* Washington, DC: National Academies Press.

Penuel, W. R., and B. J. Fishman. 2012. Large-scale science education intervention research we can use. *Journal of Research in Science Teaching* 49 (3): 281–304.

Penuel, W. R., B. J. Fishman, L. P. Gallagher, C. Korbak, and B. Lopez-Prado. 2009. Is alignment enough? Investigating the effects of state policies and professional development on science curriculum implementation. *Science Education* 93 (4): 656–677.

Penuel, W. R., B. J. Fishman, R. Yamaguchi, and L. P. Gallagher. 2007. What makes professional development effective? Strategies that foster curriculum implementation. *American Educational Research Journal* 44 (4): 921–958.

Remillard, J. T. 2005. Examining key concepts in research on teachers' use of mathematics curricula. *Review of Educational Research* 75 (2): 211–246.

Roseman, J. E., M. C. Linn, and M. Koppal. 2008. Characterizing curriculum coherence. In *Designing coherent science education. Implications for curriculum, instruction, and policy*, ed. Y. Kali, M. C. Linn, and J. E. Roseman, 13–36. New York: Teachers College Press.

Shwartz, Y., A. Weizman, D. Fortus, J. Krajcik, and B. Reiser. 2008. The IQWST experience: Using coherence as a design principle for a middle school science curriculum. *Elementary School Journal* 109 (2): 199–219.

Singer, S. R., M. L. Hilton, and H. A. Schweingruber. 2006. *America's lab report: Investigations in high school science*. ed. S. R. Singer, M. L. Hilton, and H. A. Schweingruber. Washington, DC: National Academies Press.

Spillane, J. P., and K. A. Callahan. 2002. Implementing state standards for science education: What district policy makers make of the hoopla. *Journal of Research in Science Teaching* 37 (5): 401–425.

Chapter 4

Major STEM Reforms Informing Professional Development

Richard A. Duschl

Rationale

What would science education teaching and learning look like if the design of curriculum, instruction, and assessment began by considering what we wanted students to do as opposed to what we wanted students to know? The legacy of science education in the United States has been one that for the last half century has asked, what do we want students to *know*, and what do they need to do to *know* it? An alternative perspective asks, what do we want students to *do*, and what do they need to know to *do* it? This doing perspective, in the guise of inquiry-based science education, has been with us for close to 50 years, but still we find in our curriculum, instruction, and assessment models an emphasis on *knowing* the "what" over *doing* to get at the "how" we know and why we believe.

Two findings from the recent National Research Council (NRC) summary reports *Taking Science to School* (Duschl, Schweingruber, and Shouse 2007) and *Ready, Set, Science!* (Michaels, Shouse, and Schweingruber 2008) tell us that we have underestimated the science abilities of children and misdirected our focus on teaching inquiry science. Research from learning scientists tells us that children, even before kindergarten, are more capable than we ever thought at reasoning and at doing science. Research from science studies scholars informs us that scientific inquiry is much more than conducting investigations. The data obtained from lab and field study investigations are just a starting point for "knowing and doing" practices involved in building, refining, and communicating scientific ideas and models.

A major professional development challenge facing teachers of science is learning to use assessment practices and a standards-based curriculum that will guide adaptive instruction. The incorporation of the three dimensions of the NRC's *A Framework for K–12 Science Education* (*Framework*; NRC 2012) (i.e., science practices, crosscutting concepts, and core ideas) into learning progressions reinforces the coordination of knowing and doing. The decision in the *Next Generation Science Standards* (*NGSS*; NGSS Lead States 2013) to use learning performance assessment models reinforces the importance for joining the doing and knowing practices. The alignment of curriculum-instruction-assessment models coordinated around learning progressions has the potential to organize classrooms and other learning environments around adaptive instruction (targeted feedback to students). The adoption of evidence-based designed assessment models has the potential to coordinate adaptive instruction with learning performances. The *Framework* informs the *NGSS*, the *NGSS* inform the learning goals and performances, and they in turn inform the pedagogical practices and activities that will guide teachers in conducting classroom-level assessments that help advance students' learning. Thus, a major challenge facing classroom teachers and professional development coordinators is learning how to use assessments that provide information to target students' learning strategies, which in turn guide teachers' strategies for helping provide the right feedback, for example, instruction-assisted development. The research reviews from *Taking Science to School* indicate that in science the learning strategies have several components.

Models

A Model of Learning: Science in Three-Part Harmony[1]

The role of assessment and the importance of assessment information have shifted to the classroom. We now have expectations that teachers will obtain and review students' learning performances. We distinguish achievement measures from assessment measures and respectively refer to each as "assessment of learning" and "assessment for learning." As mentioned above, the professional development challenge or focus is on assessment for learning, for instance, the classroom-level diagnostic assessments that guide teachers with monitoring and adapting the progress of learning. Our deeper understanding of children's science learning has led to the recognition that science learning involves three important domains: the *cognitive*, the *sociocultural*, and the *epistemic*. Considering these three components fundamentally broadens what we mean by science learning assessments. In the sections that follow, an overview of each domain is presented, and connections to (*Ready, Set, Science!* (*RSS*) chapters are indicated.

Cognitive Perspective

The cognitive perspective focuses (Figure 4.1) on the meaning making and knowledge formation skills students need to be proficient learners. Proficient science learners are found to use four cognitive dimensions derived from the human expertise literature (Glaser 1997) and endorsed among learning theorists (e.g., Anderson 1990; Bransford, Brown, and Cocking 1999). These are as follows: (1) structured, principled knowledge; (2) proceduralized knowledge; (3) effective problem representation; and (4) self-regulatory skills.

Taken together, the design and implementation of assessments ought to focus on integrated knowledge structures, the efficient and appropriate use of knowledge during problem solving, the ability to use and interpret different representations, and the ability to monitor and self-regulate learning and performance.

1. This section presents selected ideas from Gitomer and Duschl (2007). Interested readers are referred to the full chapter for further details on establishing multilevel coherence in assessments.

Figure 4.1

COGNITIVE DIMENSIONS

Structured, principled knowledge

Learning involves the building of knowledge structures organized on the basis of conceptual domain principles. For example, chess experts can recall far more information about a chessboard than the average person, not because of better memories but because they recognize and encode familiar game patterns as easily recalled, integrated units. [*RSS*, Chapter 3: Foundational Knowledge and Conceptual Change; Chapter 4: Organizing Science Education Around Core Concepts]

Proceduralized knowledge

Learning involves the progression from declarative states of knowledge ("I know the rules for multiplying whole numbers by fractions") to proceduralized states in which access is automated and attached to particular conditions ("I apply the rules for multiplying by fractions appropriately, with little conscious attention").

Effective problem representation

As learners gain expertise, their representations move from a focus on more superficial aspects of a problem or task to underlying deeper structures. For example, experts organize physics problems on the basis of underlying physics principles (acceleration problems), whereas novices sort the problems on the basis of surface characteristics (mass of sled, height and slope of hill). [*RSS*, Chapter 6: Making Thinking Visible: Modeling and Representation]

Self-regulatory skills

As a learner becomes aware of progress being made, or not made, with their own meaning making and understanding, they become more capable with monitoring learning and performance, allocation of time on task, and gauging task difficulty. [*RSS*, Chapter 3: Foundational Knowledge and Conceptual Change]

Sociocultural Perspective

The sociocultural perspective (Figure 4.2, p. 48) focuses on the social interactions of communicating and critiquing ideas and how these interactions influence learning. Learning from this perspective involves the adoption of sociocultural practices, in particular the science practices from the *Framework*. Students of science, for example, not only learn the content of science, they also develop an "intellective identity" as scientists by becoming acculturated to the tools, practices, and discourse of science. This perspective grows out of the work of Vygotsky and others and maintains that learning and practices develop out of social interaction.

Figure 4.2

KEY ATTRIBUTES OF SOCIOCULTURAL ASSESSMENT DESIGN

Public displays of competence

Research on productive classroom interactions indicates that public sharing and displays of student work and learning performances are important social activities. So, too, are open discussions of the criteria by which performance is evaluated as well as discussions among teachers and students about the work and dimensions of quality. Such strategies facilitate making student thinking visible through talk, argument, modeling, and representation. (See chapter Chapter 5, "Making Thinking Visible: Talk and Argument" in Michaels, Shouse, and Schweingruber 2008.)

Engagement with and application of scientific tools

A great deal of curriculum and assessment development has focused on the use of science tools and materials in conducting science investigations. Assessments ought to also include activities that require students to engage with measurement, observation, data analysis, and data representation tools of science and to reflect on the conditions that determine the applicability of specific tools and practices. (See Chapter 6, "Making Thinking Visible: Modeling and Representation" in Michaels, Shouse, and Schweingruber 2008.)

Self-assessment

A key self-regulatory skill (see Figure 4.1, p. 47) that is a marker of expertise is the ability and propensity to assess the quality of one's own work. Assessments should provide opportunities, through practice, coaching, and modeling for students to develop abilities to effectively judge their own work. (See chapter Chapter 5, "Making Thinking Visible: Talk and Argument" in Michaels, Shouse, and Schweingruber 2008.)

Access to reasoning practices

Science assessment can contribute to the establishment and development of science practice by students, facilitated by teachers. Assessments can be designed to encourage productive interactions with students that engage them in important reasoning practices regarding the building, refining, critique, and communication of models, mechanisms, theories, and explanations. (See Chapter 2, "Four Strands of Science Learning" and Chapter 7, "Learning From Science Investigation" in Michaels, Shouse, and Schweingruber 2008.)

Socially situated assessment

Expertise is often expressed in social situations (e.g., group, poster or oral presentations, written or oral critique and communication) in which individuals need to interact with others. There is often exchange, negotiation, building on others' input, contributing and reacting to feedback, and so on. Indeed, the ability to work within social settings is highly valued in work settings and insufficiently attended to in typical schooling, including assessment. (See Chapter 5, "Making Thinking Visible: Talk and Argument" in Michaels, Shouse, and Schweingruber 2008.)

Some key attributes of assessment design that would be consistent with a sociocultural perspective focus on the tools, practices, and interactions that characterize the community of scientific practice. They are (1) public displays of competence, (2) engagement with and application of scientific tools, (3) self-assessment, (4) access to reasoning practices, and (5) socially situated assessment.

Assessments exist within an educational context and can have intended and unintended consequences for instructional practice. A primary criticism of the traditional high-stakes "what we know" assessment methodology is that it supports adverse forms of instruction, for example, too much teacher talk. By attending to the sociocultural practices described above, assessment designs provide models of practice that can be used in instruction.

Epistemic Perspective

The epistemic perspective (Figure 4.3, p. 50) further clarifies what it means to learn science by situating the cognitive and sociocultural perspectives in specific scientific activities and contexts in which the growth of scientific knowledge occurs. There are two general elements in the epistemic perspective—one disciplinary, the other methodological. The first involves looking at the different knowledge-building traditions that exist for each science discipline (e.g., historical models for Earth science and mathematical models for physical science); the second considers the shared practices (e.g., modeling and measuring).

These two epistemic perspectives along with the four strands of science proficiency are not merely learning goals for students; they also set out a framework for diagnostic assessments and, in turn, the design of dynamic learning environments. The epistemic perspectives represent the knowledge and reasoning skills needed to be proficient in science and to participate in scientific communities, be they classrooms, lab groups, research teams, workplace collaborations, or democratic debates.

A Model of Assessment Systems

Bringing together the three knowledge perspectives as a learning model helps define the assessment system. The implications for an assessment system externally coherent with such an elaborated model of learning are profound. Assessments need to be designed to monitor the cognitive, sociocultural, and epistemic practices of doing science by moving beyond treating science as the accretion of knowledge to a view that science, at its core, is about acquiring data and then transforming that data first into evidence, then the evidence into patterns or models, and finally patterns or models into explanations. Enactment of these three transformations is complex and represents a professional development challenge for science teachers. The challenges and recommendations for actions can be found in my own and other contributions to the NSTA-edited volume *Everyday Assessment in the Science Classroom* (Atkin and Coffey 2003).

Figure 4.3

THE EPISTEMIC PERSPECTIVE

Disciplinary domain-specific traditions

Knowledge-building traditions in science disciplines (e.g., physical, life, Earth and space, medical, engineering), while sharing many common features (e.g., crosscutting concepts) are actually quite distinct when the tools, technologies, and theories each uses are considered. Such distinctions shape the science practices and inquiry methods adopted. For example, geological and astronomical sciences will frequently adopt historical and model-based methods to develop explanations for the formation and structures of the Earth, solar system, and universe. Causal mechanisms and generalizable explanations aligned with mathematical statements are more frequent in the physical sciences in which experiments are more readily conducted. Whereas molecular biology inquiries often use controlled experiments, population biology relies on testing models that examine observed networks of variables in their natural occurrence. (See Chapter 3, "Foundational Knowledge and Conceptual Change" and Chapter 4, "Organizing Science Education Around Core Concepts" in Michaels, Shouse, and Schweingruber 2008.)

Disciplinary domain-specific practices

The second element of the epistemic perspective includes shared practices such as modeling, measuring, and explaining that frame students' classroom investigations and inquiries. *Taking Science to School* argues that content and process are inextricably linked in science and characterized by learning that attends to the four strands of science proficiency:

1. Know, use, and interpret scientific explanations of the natural world.
2. Generate and evaluate scientific evidence and explanations.
3. Understand the nature and development of scientific knowledge.
4. Participate productively in scientific practices and discourse.

(See Chapter 2, "Four Strands of Science Learning" and Chapter 8, "A System That Supports Student Learning" in Michaels, Shouse, and Schweingruber 2008.)

As discussed above, sociocultural and epistemic perspectives about learning reshape what we mean by science understanding. The two perspectives also inject a significant and alternative justification for not only what we assess but also how we assess and who does the assessing. When the sociocultural and epistemic perspectives are included in our models of learning, it becomes clear that our standard psychometric models are markedly incomplete. Smith et al. (2006), while establishing multilevel coherence in assessment, note that "[current standards] specify the knowledge that children should have, but not practices—what children should be able to *do* with that knowledge" (Smith et al. 2006, p. 4). The *Framework* and *Taking Science to School* argue for including *practices* as demonstrations of subject matter competence, for instance, using science principles in learning performances. Assessments that ignore the practices do not adequately assess the constellation of coordinated cognitive, sociocultural, and epistemic skills that encompass science subject-matter competence. The professional development challenge facing science teachers is one that involves planning, implementing, and managing the constellation of coordinated skills. Thus, the question of whether multiple-choice assessments can adequately sample a domain is necessarily answered in the negative, for they do not require students to engage and demonstrate competence—make thinking visible—in the full set of practices of the domain.

What might an assessment design that does account for the sociocultural and epistemic perspectives look like? First, consider the essential features of scientific inquiry that appear in Figure 4.4. Next think of the essential features of inquiry found in *Inquiry and the National Science Education Standards* (NRC 2001) as an evidence-explanation (E-E) framework containing three transformations (Duschl 2003):

1. data to evidence, or determining which measurements and observations to use, to count as evidence
2. evidence to patterns, or searching the evidence for trends, cycles, and models that suggest patterns
3. patterns to explanations, or developing or locating explanations on the basis of the evidence or models selected

Figure 4.4

ESSENTIAL FEATURES OF SCIENTIFIC INQUIRY

Learners are engaged by scientifically oriented questions.

Learners give priority to evidence, which allows them to develop and evaluate explanations that address scientifically oriented questions.

Learners formulate explanations from evidence to address scientifically oriented questions.

Learners evaluate their explanations in light of alternative explanations, particularly those reflecting scientific understanding.

Learners communicate and justify their proposed explanations.

Source: NRC 2001, p. 25.

Teachers' assessments of student inquiry and science practices should focus on students progressing across these three transformations. Students are asked to make reasoned judgments and decisions (e.g., arguments) during three critical transformations in the E-E framework: *selecting* data to be used as evidence; *analyzing* evidence to extract or generate models or patterns of evidence; and *determining and evaluating* scientific explanations to account for models and patterns of evidence.

Unpacking these transformations brings us to the central role practices have in science learning. Now consider the *Framework*'s eight science and engineering practices presented in Figure 4.5. What the inquiry transformations and science and engineering practices signal is the need to make evidence problematic for students—not something given but rather something that is obtained, evaluated, wrestled with, argued over, applied, represented, and communicated. In this way, the three-step E-E approach emphasizes the progression of "data-texts" (e.g., measurements to data to evidence to models to explanations) found in science, and it embraces the aforementioned cognitive, sociocultural, and epistemic perspectives.

Teaching with a focus on scientific inquiry and practices requires much more than providing and managing materials and activities for students to conduct investigations. Engaging students in kit-based science or lab investigations in and of itself is not inquiry. The kind of instruction and assessment we need to consider is that which has students examining how we know what we know and why we believe this knowledge over other competing knowledge claims. Such instruction is grounded in considering what counts as evidence and as explanations and involves students in the critique and communication of the same.

Figure 4.5

SCIENTIFIC AND ENGINEERING PRACTICES FROM THE *FRAMEWORK*

- Asking questions (for science) and defining problems (for engineering)
- Developing and using models
- Planning and carrying out investigations
- Analyzing and interpreting data
- Using mathematics, information and computer technology, and computational thinking
- Constructing explanations (for science) and designing solutions (for engineering)
- Engaging in argument from evidence
- Obtaining, evaluating, and communicating information

Source: NRC 2012, p. 3.

In the E-E framework, scientific and engineering practices 3 to 6 are highlighted. Think of the transformations as stepping-stones to guide instruction-assisted development. In this case, the focus is on students' opportunities to examine the development of data texts. During each transformation, students are encouraged to share their thinking by engaging in argument, representation and communication, and modeling and theorizing. Teachers are guided to engage in assessments by comparing and contrasting student responses to each other and, importantly, to the instructional aims, knowledge structures, and goals of the science unit. Examination of students' knowledge, representations, reasoning, and decision making across the transformations provides a rich context for conducting assessments.

The advantage of this approach resides in the formative assessment opportunities for students and the cognitive, sociocultural, and epistemic practices that comprise "doing science" that teachers will monitor. Two recent reports (Duschl, Schweingruber, and Shouse 2007; NRC 2012) offer insights into the challenge of designing assessments that incorporate these additional perspectives and practices. Both reports introduce ideas about "learning progressions" and "learning performances" as strategies to rein in the overwhelming number of science standards and benchmarks. The new *Framework* agenda is to focus on core ideas (e.g., deep time, atomic molecular theory, evolution) in concert with practices and crosscutting concepts that ought to be at the heart of science curriculum sequences.

Lessons Learned

The professional development challenge is simple to state but complex to implement and manage. Teachers need to develop new ways of listening to and monitoring students' scientific reasoning and thinking. And this requires that classrooms and other learning environments be reformed. One key component in the redesign of learning environments recommended in *Taking Science to School* is developing extended instructional contexts, for instance, teaching sequences and learning progressions (see Duschl, Maeng, and Sezen 2011 for a review of this emerging field of education research). A second redesign feature is embedding formative assessment tasks and activities into the instructional sequences. Such tasks and activities help make students' thinking visible.

This gives rise to two additional teacher challenges: One teacher challenge is developing students' abilities to communicate and represent what they know and how and why they believe what they know. The other teacher challenge is coordinating the review and analysis of students' ideas and then incorporating these ideas back into the lessons. As stated above, good teachers have always understood the importance of providing feedback. With the new models of learning emphasizing practices and using knowledge, new more complex forms and processes of feedback are required. Importantly, in the redesigned learning environments these don't always emerge from the teachers but also from students' peers and the students themselves. As students become more adept at communicating and representing scientific information, teachers will, in turn, be in a better position to listen and give feedback. It is a symbiotic relationship.

Reflections for the Future

Assessing Learning With Learning Performances

The *Framework*'s three dimensions represent a more integrated view of science learning that should reflect and encourage science activity that approximates the practices of scientists. Let's consider what that means for teaching and assessing the crosscutting concepts. One consideration is that assessment tasks should be cumulative across a grade band and include many of the sociocultural and epistemic perspectives that are a part of doing science, for example, talk and arguments, modeling and representations. The crosscutting concept assessments would be less frequent; each term or annually there would be a performance assessment task that would reveal how students are enacting and using the three dimensions. The majority of crosscutting concept assessment tasks will be constructed response and performance assessments. If the goal is to gauge students' crosscutting concept enactments when asked to ascertain patterns, generate mechanisms and explanations, distinguish between stability and change, provide scale representations, model data, and otherwise engage in various aspects of science practices, then the students must show evidence of "doing" science and of critiquing and communicating what was done.

The message from the *Framework* is that there are important interconnections between crosscutting concepts and disciplinary core ideas. "Students' understandings of these crosscutting concepts should be reinforced by repeated use in the context of instruction in the disciplinary core ideas ... the crosscutting concepts can provide a connective structure that supports students' understanding of sciences as disciplines and that facilitates their comprehension of the systems under study in particular disciplines" (p. 101). What this says is that the crosscutting concepts are to be embedded within or conjoined across coherent sequences of science instruction. The *Framework*'s three dimensions—science practices, crosscutting concepts, core ideas—send a clear message that science learning and instruction must not separate the knowing (concepts, ideas) from the doing (practices). Thus, the assessment strategies teachers adopt for pupils' understandings of and enactments with the seven crosscutting concepts must also conjoin the knowing and doing.

The *Framework* and the *NGSS* provide teachers with an agreed upon set of curricular goals. The *NGSS* are presented in a learning performances format. That is, the conceptual knowledge to be understood is conjoined with a practice that frames the use of the knowledge. For example, consider a task that asks students to explain how an odor travels through a room. It could be assessed using the grade band information found in the *Framework*'s crosscutting concept Energy and Matter: Flow, Cycles, and Conservation (Figure 4.6).

The expectation is for students to use some conceptual knowledge (e.g., states of matter, atomic theory) with a practice (e.g., modeling, asking questions) to develop a mechanism (gas or particle diffusion) that explains the odor's movement. What a teacher is seeking is evidence that students are developing a model of matter made of particles. Related tasks could be mechanisms for the diffusion of a colored dye in water, the separation of sediments in water, or the role of limiting factors in an ecosystem or chemical reaction. The tasks can be gathered over the grade band to develop a portfolio of evidence about students' crosscutting concept understandings and enactments.

Summary

The inclusion of crosscutting concepts and the focus on inquiry practices in the *Framework* continues a 50-year history in U.S. science education that both scientific knowledge and knowledge about science are important K–12 education goals. It's the dual agenda for science. The crosscutting conceptss are best thought of as the learning goals for science literacy. But learners' success hinges on doing the science and getting quality feedback from teachers and peers. The coordination of the three *Framework* dimensions reinforces the importance of not separating the doing from the knowing. The alignment of curriculum-instruction-assessment models that is coordinated around learning progressions has

Figure 4.6

CROSSCUTTING CONCEPTS FOR ENERGY AND MATTER: FLOW, CYCLES, AND CONSERVATION

Energy and matter: Flows, cycles, and conservation. Tracking fluxes of energy and matter into, out of, and within systems helps one understand the systems' possibilities and limitations.

K–2 Focus is on basic attributes of matter in examining life and Earth systems. Energy is not developed at all at this grade band.

3–5 Macroscopic properties and states of matter, matter flows, and cycles are tracked only in terms of the weights of substances before and after a process occurs. Energy is introduced in general terms only.

6–8 Introduce role of energy transfers with flow of matter. Mass/weight distinctions and idea of atoms and their conservation are taught. Core ideas of matter and energy inform examining systems in life science, Earth and space science, and engineering contexts.

9–12 Fully develop energy transfers. Introduce nuclear substructure and conservation laws for nuclear processes.

Source: NRC 2012, pp. 94–96.

great potential to organize classrooms and other learning environments around adaptive instruction (targeted feedback to students) and instruction-assisted development. We considered a learning model based on cognitive, sociocultural, and epistemic perspectives. We considered how in science education one important domain is assessing inquiry—monitoring and giving feedback on learners' conceptual meaning making as well as on the investigative, communicative, and reasoning science practices that make up the essence of doing science. Although science involves doing investigations to obtain evidence, science is so much more. Learning science is not simply learning facts and knowledge claims. Learning science is using facts and claims to pose, build, and refine explanatory models and mechanisms. Students need targeted feedback from teachers and peers to accomplish such building and refining practices.

Taking Science to School tells us we need to move from general science processes to domain-specific science practices. Measurement and observation practices in the physical sciences are very different practices in the life and Earth sciences. Curriculum materials and instructional methods need to change from thinking about teaching general inquiry processes to thinking about teaching and having students engage in specific science practices. This is what the four strands of science proficiency in *RSS* are all about:

- Strand 1: *Understanding scientific explanations*, e.g., understanding central concepts and using them to build and critique scientific arguments
- Strand 2: *Generating scientific evidence*, e.g., generating and evaluating evidence as part of building and refining models and explanations of the natural world

- Strand 3: *Reflecting on scientific knowledge*, e.g., understanding that doing science entails searching for core explanations and the connections between them
- Strand 4: *Participating productively in science*, e.g., understanding the appropriate norms in presenting scientific arguments and evidence and to practice productive social interactions with peers around classroom science investigations (Michaels, Shouse, and Schweingruber 2008)

So what does research tell us inquiry sounds like?

- Planning investigations for data is children selecting questions, tools, and schedules for observation and units for measurement.
- Data collection to evidence is children observing systematically, measuring accurately, structuring data, and setting standards for quality control.
- Evidence to searching for patterns and building models is children constructing and defending arguments, presenting evidence, engaging in mathematical modeling, and using physical and computational tools.
- Patterns and models to generate explanations is children posing theories, building and reporting conceptual-based models, considering alternatives, and generating new productive questions.

Fundamentally assessing inquiry is all about making thinking visible through talk and argument and through modeling and representation, Chapters 5 and 6, respectively, in *RSS*. Professional development for implementation of the *Framework* and the *NGSS* begins with teachers and students learning how to listen for and respond to the core practices and crosscutting concepts involved with the critique and communication of science claims. In science over the last century, we have learned how to learn about nature. In education over the last century, we have learned how to learn about learning. As we proceed into the 21st century, let us learn how to meld together these two endeavors. The *Framework* and the *NGSS* are a great beginning, but successful implementation will only come about through the participation and commitment of teachers. The set of research and policy reports from the NRC are strong evidence that we as teachers, science educators, and science education researchers have learned how to learn about learning. The teacher is the key that will help us unlock how to fully understand the best coherent sequences for learning and teaching.

References

American Association for the Advancement of Science (AAAS). 1989. *Science for all Americans*. New York: Oxford University Press.

Anderson, J. R. 1990. *The adaptive character of thought*. Hillsdale, NJ: Erlbaum.

Atkin, J. M., and J. Coffey, eds. 2003. *Everyday assessment in the science classroom*. Arlington, VA: NSTA Press.

Bransford, J., A. Brown, and R. Cocking, eds. 1999. *How people learn: Brain, mind, experience, and school*. Washington, DC: National Academies Press.

College Board. 2009. Science: College Board standards for college success. *professionals.collegeboard.com/profdownload/cbscs-science-standards-2009.pdf.*

Duschl, R. 2003. Assessment of inquiry. In *Everyday assessment in the science classroom*, ed. J. M. Atkin and J. Coffey, 41–59. Arlington, VA: NSTA Press.

Duschl, R. 2012. The second dimension—crosscutting concepts: Understanding A Framework for K–12 Science Education. *Science and Children* 49 (6): 10–14.

Duschl, R., S. Maeng, and A. Sezen. 2011. Learning progressions and teaching sequences: A review and analysis. *Studies in Science Education* 47 (2): 123–182.

Duschl, R. A., H. A. Schweingruber, and A. W. Shouse, eds. 2007. *Taking science to school: Learning and teaching science in grades K–8*. Washington, DC: National Academies Press.

Gitomer, D. H., and R. Duschl. 2007. Establishing multi-level coherence in assessment. In *Evidence and decision making: The 106th yearbook of the National Society for the Study of Education, Part 1*, ed. P. Moss, 288–320. Chicago, IL: National Society for the Study of Education.

Glaser, R. 1997. *Assessment and education: Access and achievement.* CSE tech. rep. 435. Los Angeles, CA: National Center for Research on Evaluation, Standards, and Student Testing.

Michaels, S., A. W. Shouse, and H. A. Schweingruber, eds. 2008. *Ready, set, science!: Putting research to work in K–8 science classrooms*. Washington, DC: The National Academies Press.

National Assessment Governing Board (NAGB), U.S. Department of Education. 2008. NAEP 2009 science framework development: Issues and recommendations. *www.nagb.org/content/nagb/assets/documents/publications/frameworks/science-09.pdf.*

National Research Council (NRC). 2001. *Inquiry and the national science education standards.* Washington, DC: National Academies Press.

National Research Council (NRC). 2012. *A framework for K–12 science education: Practices, crosscutting concepts, and core ideas.* Washington, DC: National Academies Press.

NGSS Lead States. 2013. *Next Generation Science Standards: For states, by states*. Washington, DC: National Academies Press. *www.nextgenscience.org/next-generation-science-standards.*

Smith, C., M. Wiser, C. Anderson, and J. Krajcik. 2006. Implications of research on children's learning for assessment: Matter and atomic molecular theory. *Measurement: Interdisciplinary Research and Perspectives* 4 (1–2): 11–98.

Part II

State and District Models and Approaches

Chapter 5

Using Constructivist Principles in Professional Development for STEM Educators: What the Masters Have Helped Us Learn

Karen Charles

Rationale

In the field of science, technology, engineering, and mathematics (STEM) education, the transition from classroom teacher to professional developer is not uncommon. What is uncommon is that any type of support might accompany this transition. Most teachers have not encountered professional development classes in their education coursework, nor is there an abundance of inservice workshops dedicated to training professional developers. This raises the question, where does one learn this craft and find the follow-up support necessary to sustain ongoing improvement? Further, how might such programs be designed to help professional developers deepen their content and pedagogical knowledge regarding teaching and learning science and mathematics?

Model

These were the questions that the federally funded Eisenhower Consortium at the southeastern regional educational lab SERVE intended to answer by creating the Technical Assistance Academy for Mathematics and Science Services that hosted staff developers from a six-state region from 1996–2005. At that time, standards-based reform documents (e.g., NRC 1996; AAAS 1989; NCTM 1989, 1991) were calling for schools and teachers to replace the practice of transmitting knowledge and facts with active engagement to promote deep understanding, critical thinking, and authentic learning. This approach, teaching for understanding (McLaughlin and Talbert 1993), required teachers to establish a classroom culture in which knowledge is developed collaboratively, facts are challenged continually in discourse, and both teachers and students engage in learning and inquiry. These new roles for teachers and learners required developing new support systems for instructors that encouraged professional growth and change.

The challenge was, and still is, to promote constructivist classroom practices in a professional development experience that is itself constructivist. Teachers do not need to endure lectures about constructivist learning situations; they need to experience it themselves, question their own beliefs, and confront their own disequilibrium. Staff developers need to know how to design these environments for their professional development offerings. Their workshops and seminars need to reflect the change they are promoting.

Background

Throughout the last century, passive learning for both adults and children was criticized by renowned educational leaders (e.g., Dewey 1916; Bruner 1971; Piaget 1973, 1974; Friere and Schor 1987). In evaluating the federal technical assistance programs of the day, Turnbull (1995) wrote:

> Because real school improvement requires teachers and other educators to learn, assistance providers will have to do things that help learning take place: create—and help people use—opportunities for problem solving, professional conversations, and reflection. Such an assistance agenda exceeds the current policy expectation that telling and showing are adequate means of providing help. (Turnbull 1995, p. 12)

The challenge in professional development, particularly for STEM educators, is to replace passiveness with experiences that more closely reflect the ideals of inquiry and constructivist learning and to avoid the pitfall of "telling and showing" that has been accepted as reasonable professional development. The questions then become, does attending to constructivist principles when designing professional development experiences for professional developers contribute to building capacity, shifting attitudes, and improving practices? Can professional developers be lured away from the traditional and easy approach (telling and showing) and embrace the more open and unpredictable principles of constructivism?

Sparks and Loucks-Horsley (1989) indicated that traditional professional models have not always successfully engaged teachers in the type of learning that translates into classroom change, and structures that support the engagement of teachers in inquiry and reflection for the purpose of promoting improvement were difficult to find. They acknowledged the need for more descriptive studies that focus on the relationship between professional development and the professional growth of educators and the leadership roles they fill.

At the time, however, the problem was that the literature on constructivism in science and mathematics focused mainly on the effectiveness of constructivist-related instructional approaches in helping K–12 students progress in understanding. Through the Technical Assistance Academy, participants experienced inquiry, discourse, and the active engagement called for by national standards of the time and called for again today by the *Next Generation Science Standards* (NGSS Lead States 2013). The participants, who were all professional developers, were then given the time and opportunity to reflect on the application of these experiences to their own practices. The academy hoped to contribute to the literature by studying the effectiveness of a constructivist learning environment in helping professional developers promote inquiry and problem solving by designing experiences for other educators that modeled these principles.

Teaching for Transmittal

When teaching for transmittal is practiced, knowledge becomes the information that passes from the mouth of the teacher to the notes of the student without going through the minds of either. In this environment, creativity is not nurtured, critical thought does not develop, and students fail to make connections that would encourage them to see new uses for old information and cultivate a fluency of ideas and flexibility of focus. Students cannot develop critical-thinking skills when they are not required to inquire, question, and challenge. Teaching for transmittal is safe. By contrast, when teachers pose problems and allow students to do so as well, they relinquish control—and when they relinquish control, they open the door for the construction of knowledge.

Overview of Constructivist Thought

Constructivism holds that one learns as the result of a direct interaction with concepts, the struggle that accompanies this interaction, and the deep reflection that guides these two experiences. These ideas formulate a constructivist model of learning that can be explained as a cycling of four types of activities that help people make meaning out of their experiences: (1) equilibration: attempting to establish or reestablish a balance between new information and previous ideas; (2) reflection: considering the impact of the new state of equilibrium; (3) construction: attempting to incorporate information provided by stimuli into an existing system of knowledge; and (4) accommodation or assimilation: applying the new construct to previously held beliefs in search of acceptance or potential conflict. The rotation of this constructivist cycle formed the framework of the academy with activities and experiences specifically chosen to require equilibration and, ultimately, a new state of equilibrium. Think of Le Chatelier's principle—that if change is introduced to a system in chemical equilibrium then the system will find a new equilibrium—with people rather than molecules.

Constructivist Instructional Practices

Constructivist instructional practices are compelling in that they promote inclusion and participatory learning, and they honor the learner in the process of learning. As a learning theory tracing its roots to Piaget (1973), constructivism has recognizable components, ideology, and tenets. Interpretations of Piaget's ideas are found in science and mathematics education reform documents as well as in newer models of professional development. If teachers are to create classroom climates that promote deep understanding, critical thinking, and authentic learning, in which learners are actively engaged in exploring, conjecturing, and constructing their own knowledge, then they too must learn in an environment in which they can build the capacity to confront new situations with internalized strategies for adoption (assimilation) or adaptation (accommodation).

The challenge for professional developers is to design and deliver opportunities for teachers in which the facilitator models the role of the instructor in the classroom. The teachers become the students, the facilitator becomes the teacher, and disequilibrium abounds. Teachers confront their beliefs about how students learn, why questions are often the best answers, and when to let students guide the lessons. This type of environment supports Piaget's belief that knowledge results from an interaction between the stimulus and the learner, is constructed by the learner, and cannot be provided by the teacher (Bringuier 1980). Expressing these same ideas earlier, Dewey (1916) wrote:

> No thought, no idea can possibly be conveyed as an idea from one person to another. When it is told, it is, to the person to whom it is told, another fact, not an idea. Only by wrestling with the conditions of the problem at first hand, seeking and finding his own way out, does he think. (p. 84)

This implies that just as teachers need to design and situate learning experiences in the context of students' prior knowledge, skills, beliefs, and values, so must professional developers for their adult audiences. The developers need to view their participants as a community of learners wherein the educators are supported in taking risks, testing new ideas, and making sustained efforts at serious learning in an atmosphere of mutual trust (McLaughlin and Talbert 1993). In this setting, teaching becomes more difficult than learning (Heidegger 1968) because the professional developer needs to be willing to step aside to let learning occur and to allow teachers to learn in the same manner as is advocated for their students.

Professional Development

As a field of study with its own content and pedagogy, professional development began emerging in the 1970s and solidified in the mid- to late-'80s (Killion and Harrison 1997). Almost as quickly as it emerged, pioneers in the field challenged the earliest held tenets and proposed radical departures from commonly held beliefs and practices about how adults learn best, what approaches ensure internalization of new material, the role of modeling, and the role of reflection. The "sage on the stage" was being challenged by the "guide on the side," who began refusing the role of one-day change agent.

New thinking suggested that professional developers become problem makers and problem solvers and that they help their clients construct meaning out of their professional development experiences. It followed that they should learn their craft in the same way, through the disequilibrium and adaptation that follows. Fosnot (1996a) and Killion and Harrison (1997) suggested that professional developers need to focus on their own professional development, assessing their own skills and monitoring how they respond to challenges. This recommended shifting paradigm closely parallels the principles of inquiry and constructivism, which suggest that self-awareness and reflection promote learning (Greene 1978; Piaget 1974).

Effective Professional Development Strategies

Darling-Hammond and McLaughlin (1996) proposed that effective professional development allows educators to struggle with the uncertainties that are inherent in both the role of the instructor and the role of the learner. This idea of struggle is consistent with experiencing the assimilation and accommodation that accompany the construction of knowledge. Duke (1990) indicated that several components critical to the development of effective professional development experiences for educators need to be present in order for accommodation to occur: (1) an awareness that there is a need for change by examining assumptions and reading challenging material; (2) time to learn; and (3) a variety of points of view to stimulate reflection, experimentation, and collegiality. This perspective represents a major shift from the large group, one-hour, after-school workshops that many teachers have endured. Duke (1993) offered insight into the relationship between the accommodation in constructivist learning and effective professional development:

> Professional growth involves learning, but it is more than learning. While learning may represent the acquisition of new knowledge, growth implies the transformation of knowledge into the development of the individual. Growth is a qualitative change, movement to a new level of understanding, the realization of a sense of efficacy not previously enjoyed … Adults learn all the time, but growth, particularly professional growth, is rarer. As teachers gain experience, they may perceive less need to grow. New knowledge is increasingly filtered through well-informed cognitive structures, with the result that dissonant information is often excluded or discredited. Only knowledge that confirms prior beliefs and assumptions tends to be absorbed. (p. 703)

When professional developers monitor their own growth and continue to expand their skills, they are no longer satisfied with replicating the workshops, training sessions, and ideas of their mentors. Instead, they transform, adapt, and modify new ideas to create new materials and experiences for their audiences. They pass along these new constructs to others in the hope that even more new constructs will emerge from their participants. Replication, fidelity to a model, is no longer the goal; re-creation, or adaptation of a model, is the prize. This is the true measure of building capacity.

Adult Learners

Knowles (1978) offered a theory of adult learning that still underpins most of the current professional development premises: (1) As adults mature, they tend to prefer self-direction; (2) adult experiences are a rich resource for learning, and adults learn from these experiences through experiential techniques such as discussion; (3) adults are aware of specific learning needs generated by real-life events; and (4) adults are competency-based learners who want to apply their learning to their immediate circumstances. Adults like new information to mesh with old ideas (assimilation) or need time to reflect if new information conflicts with previously held beliefs (accommodation). Changing people's long-held beliefs takes time and careful intervention in an environment that allows for both. Adults enjoy working in small groups and practicing new skills to increase confidence. They prefer a blend of active and passive, serious and whimsical to keep their attention.

Implications for Professional Development

An overview of the implications of constructivism for professional developers can be extracted from the work of a number of researchers writing in the 1990s and interpreting and applying the ideas of the early pioneers of educational thought (e.g., Fosnot 1996a, 1996b, 1996c; von Glasersfeld 1996; Julyan and Duckworth 1996; Holzer 1994). In her research, Fosnot (1996c) posited that learning is developmental, disequilibrium facilitates learning, reflection is a driving force of learning, and learning leads to the development of "big ideas" that can be generalized across experiences. She felt that dialogue provokes thinking, and that newly constructed ideas should be shared with and communicated to one's professional community.

> If understanding the teaching/learning process from the constructivist view is itself constructed, and if teachers tend to teach as they were taught, rather than as they were taught to teach, then teacher education needs to begin with these traditional beliefs and subsequently challenge them through activity, reflection, and discourse. (Fosnot 1996c, p. 206)

> Just as young learners construct, so, too, do teachers. Teacher education programs based on a constructivist view of learning need to do more than offer a constructivist perspective in a course (or workshop) or two. Teachers' beliefs need to be illuminated, discussed, and challenged. Only through extensive questioning, reflecting, and constructing will the paradigm shift in education—constructivism—occur. (Fosnot 1996c, p. 216)

Julyan and Duckworth (1996) suggested that teachers need to pose ideas that are interesting and worthy of time and attention, listen to learners' interpretations and confusions and honor the state of "not knowing," and maintain an atmosphere of playfulness to release frustrations.

> It is the learner alone who makes the connections in any meaningful way, and it is these connections that should be of interest to the teacher.... Constructing understanding requires that the students have opportunities to articulate their ideas, to test ideas through experimentation and conversation, and to consider connections between the phenomena that they are examining and other aspects of their lives. (Julyan and Duckworth 1996, pp. 56, 58)

Constructivism is not complex, but it defies Rousseau's production model of attaining knowledge by pouring it in (Martin 1981). It requires effort, innovation, patience, and an unwavering belief that teaching and learning should be synonymous. Effort must be focused on the learner's experience, how the learner typically processes information, and the frequency and intensity of stimuli necessary for the learner to experience disequilibrium. For innovation to occur, teachers must be comfortable enough with their discipline to create meaningful experiences to engage their students. Teachers need to possess deep content and pedagogical knowledge, routinely recognize real-world situations and applications that could serve as platforms for the presentation of information, create the means for maintaining their students' interest and attention, and continually assess student understanding for the purpose of extending their own understanding and planning instruction that meets the needs of all students. Likewise, professional developers need to provide sufficient time for participants' ongoing reflection (i.e., struggle, construct and deconstruct, accept and reject, disequilibrate and equilibrate, and ultimately learn).

The literature on constructivism indicates that the struggle and the reflection it spawns are critical to learning (Piaget 1973). As is evident from the preceding discussion, the classic literature of professional development assures us that if change is to occur in American classrooms, teachers need professional development experiences that provide them with time and opportunities to struggle with ideas and to reflect, colleagues with whom to debate, and a safe environment in which to experiment. To determine whether constructivist principles were transmitted effectively in the design and delivery of the Technical Assistance Academy for Mathematics and Science Services, this study applied the principles of the Concerns-Based Adoption Model (CBAM; Hall and Hord 1987) to the data gathered.

CBAM

In its infancy, the field of professional development offered programs in which experts presented techniques, teachers listened politely and returned to their classrooms to conduct business as usual, and professional development leaders tolerated "sort of implementation" (Champion 1998, p. 7). Hall and Hord (1987) found that funding and time committed to professional development were challenged because of the lack of evidence that teachers' practices changed as a result of new information. They further found that sound educational research was not translating into effective classroom strategies because the new information was not transferring. This finding is consistent with the adaptive process in constructivism: one can block assimilation and accommodation by choosing to ignore the new information and, thus, change nothing.

Hall and Hord (1987) studied professional developers who responded by shifting from the traditional practices that had defined their discipline to the more innovative practices that were being encouraged. These new strategies included modeling rather than lecturing, actively engaging participants in the learning process, allowing time for practicing new skills, reflecting on the innovation, and providing follow-up to the session to support the transfer of the intended change and to respond to the concerns of the teachers. Hall and Hord gauged the shift displayed by the professional developers they studied using a tool they developed to support the CBAM framework, the Levels of Use indicator. CBAM emerged from 10 years of research and provides a common language for describing implementation efforts and identifying behavioral patterns to anticipate when change is expected or desired.

Levels of Use

Early research in the Levels of Use dimension of CBAM indicated that the use of an innovation (or new practice) could be determined beyond a simple yes/no level of response. Hall and Hord (1987) observed, identified, and operationally defined eight specific behaviors in order to explain a hierarchical pattern associated with the use of an innovation (Table 5.1). The focus was not on whether one used a new skill but to what extent. Behaviors in the eight levels can be thought to parallel constructivist struggles that occur when disequilibrium results from a stimulus, and one must make meaning through accommodation rather than assimilation. In this case, the struggle to incorporate new information can result in movement to a higher level of use. Time and opportunity to reflect are helpful in guiding individuals through the Levels of Use just as they are helpful in reestablishing equilibrium.

The Levels of Use framework is especially helpful in documenting the behavioral dimensions of change (i.e., the extent to which teachers report and can be observed implementing a new practice). Three of the eight levels in this model describe degrees of nonuse and five describe levels of use. The nonuse levels include nonuse (no action is occurring), orientation (one is seeking information), and preparation (an individual has decided to try something new and is preparing to do so). The five stages of use indicate a gradient experience as one begins to implement a new idea, rather than implying mastery or comfort regarding a new routine. The stages of use are mechanical (self-consciously following a format without deviation), routine (comfortably adopting a new pattern of practice), refinement (creating and testing modifications to the original), integration (informing and encouraging others to try the innovation), and renewal (seeking new information to supplement the innovation).

The three stages of nonuse allow for the fact that people slowly move into an understanding of or confrontation with new information. They may be unaware (level 0), interested enough to learn more (level 1), or acquiring the materials and information needed to try something new (level 2). Individuals in the Use phase can either progress through a number of the stages or operate at one level with no further growth. At level 6, renewal, the learner has become the teacher—capacity has been built in the learner to such an extent that the teacher and the learner are now one and the same.

The Technical Assistance Academy

The Technical Assistance Academy was designed as a professional development institute for professional developers in which participants were introduced to current research, emerging resources, and facilitation techniques; provided tools for extending their competencies as professional developers; and given time to reflect and examine their own attitudes and beliefs. It was created to reflect and model what should be happening in mathematics and science education workshops and classrooms. The goal of the academy was to give professional developers an ongoing experience intended to improve their planning and delivery of professional development programs in an environment steeped in constructivist strategies.

Format

The academy met for four days, twice a year for two years (phase 1), and once a year for the next three years (phase 2). Participants committed to the first two years and had the opportunity to opt for continued services for the next three years. Forty-five of the original 73 phase 1 participants selected this option. All participants were supported throughout the year with a variety of technical assistance services from the Eisenhower Consortium staff, including site visits; classroom observations; lesson development; presentations for colleagues of academy participants; collaboration on the design of professional development sessions; support for attendance at national conferences; and procurement of facilitators, materials, and resources.

Resources

The major curriculum resource for the academy was *Facilitating Systemic Change in Science and Mathematics Education: A Toolkit for Professional Developers* (Regional Educational Laboratory Network 1995). Materials and activities in the toolkit focus on initiating, planning, and managing change as related to science and mathematics education. Incorporating constructivist thinking with the premise of the toolkit, the academy was designed based on the following principles: everyone can learn; learning requires active participation in meaningful tasks; participants have a wealth of expertise to acknowledge and tap; community building is important for learners; and questioning, problem solving, and reflection are critical to learning.

Table 5.1 Behaviors associated with the Levels of Use

Level	Behavior
	Nonuse
0-Nonuse	User has little or no knowledge of the innovation and is doing nothing to get involved with it.
1-Orientation	User is acquiring information about the innovation and is exploring its value to and demands upon the user.
2-Preparation	User is preparing for first use of the innovation.
	Use
3-Mechanical	User is focused on the short-term effect of the innovation with benefit to the user being the key interest. Innovation is replicated, not adapted; delivery is awkward.
4a-Routine	Delivery is smooth, but no changes or improvements to the innovation are evident.
4b-Refinement	User is varying the innovation to meet the needs of the recipients.
5-Integration	User is extending the innovation to colleagues to benefit others.
6-Renewal	User modifies the innovation extensively and seeks additional information to incorporate when creating alternatives to the original.

Source: Hall and Hord 1987.

Sessions

The Eisenhower Consortium staff designed the first academy session but responded to participants' questions, needs, and recommendations and involved them in the design of the next three sessions of Phase 1 (Table 5.2). The activities for session 1 pushed participants to construct meaning by examining their prior knowledge in light of their new experiences. Specifically, they struggled to decide whether change or improvement meant refining a current skill; to rethink what is being done and how it is being done; to reexamine goals and outcomes in light of new ideas, approaches, and strategies; or to abandon current, yet comfortable, skills in favor of innovation and risk taking.

In session 2, participants explored the seminal work of Mary Budd Rowe (1978) and were surprised to realize that their own particular preferences (e.g. content-, process-, or discovery-focused instruction) could bias the design of their activities and workshops. This "ah-ha" moment was intentionally built into the academy activity to promote the personal reflection on one's own approach to curriculum delivery that Rowe felt was critical to the learners' success:

> Curriculum is not a static thing. We change our conceptions of what content and emphasis should prevail as we accumulate experience and try to foresee the kinds of futures in which our children will have to operate. Will what we do develop sufficient flexibility and sufficient mental and emotional fluency to put them in command of their fates rather than make them victims of circumstance? (Rowe 1978, p. 23)

Table 5.2 Focus of phase 1 academy sessions

Session	Theme
Session 1	*Understanding change* Develop a collaborative community, a climate for change; provide activities that promote the construction of new ideas through reflective thought
Session 2	*Curriculum delivery* Ground curriculum delivery practices in the literature and principles of educational leaders; study the national content standards
Session 3	*Workshop design* Study and apply the principles of adult learning and constructivism to the development of engaging professional development activities
Session 4	*Facilitation skills* Demonstrate emerging design techniques and presentation skills through the delivery of original activities

Proust (1934) wrote that the real art of discovery consists not in finding new lands but in seeing with new eyes. During session 3, participants reflected on whether the academy had prompted them to see their roles and responsibilities differently. They indicated that improving their workshop design skills was critical to their professional improvement; and, to meet this need, session facilitators "unpacked" their presentations for participants, describing the purpose and placement of each element, each question, and each activity. Participants were beginning to ask "why" instead of "what" with regard to professional development design. They were seeing their role as professional developers with new eyes.

In session 4, participants showcased their emerging professional development skills and celebrated their successes. Participants worked across state lines to design model lessons and workshops on topics such as team building and equity, and Eisenhower Consortium staff provided feedback throughout the creation and delivery of the sessions. Participants also provided self-report data, reflecting on what they were doing differently and how their professional experiences for their clients had improved.

Lessons Learned

Data Analysis and Results

One year after the conclusion of phase 1, the 45 educators who continued to participate in the academy were interviewed to assess their use of the resources and skills developed through participation in the academy, experiences related to the academy goal of capacity building, and needs for further technical assistance from the Eisenhower Consortium staff. The interview data, coupled with site visits, classroom or workshop observations, technical assistance, and survey and report data, provided enough robust information for evaluators to determine the level of use of new skills and strategies by these 45 participants.

Data concerning participants' insights related to the constructivist strategies (Fosnot 1996) they experienced during the academy were analyzed using descriptive statistics and content analysis procedures. Data concerning participant's reported and observed use of new skills and resources were analyzed using the Levels of Use framework of CBAM to document the extent of change evident in each participant's practice.[1]

Participants' Insights on the Constructivist Approach to Academy Design

A primary goal of the evaluation of the academy was to determine to what extent participants incorporated skills and knowledge from the academy into their daily practice and to relate these successes to the constructivist approaches that defined the academy. Self-report data from the final evaluation indicated that 71% or more of the participants reported that the academy affected their work (Table 5.3, p. 72), and 93% or more reported a moderate or major impact on their agency (Table 5.4, p. 72).

1. More detailed methodological information is available from the author.

Table 5.3 Impact of the academy on participants' work (n = 45)

Impact on participant's work	n (%)
It confirmed my current practices.	33 (73)
I have considered doing something new in my job.	32 (71)
I have started to try something I learned at the academy.	40 (88)
I have incorporated something I learned into my job.	43 (96)
I have shared ideas with a colleague.	42 (91)
The academy has not had an effect on how I do my job.	0 (0)

Table 5.4 Impact of the academy on participants' agency (n = 45)

	Level of impact, n (%)			
Impact on participant's agency	Major	Moderate	Small	None
Generated awareness of new information	32 (71)	12 (27)	1 (2)	0 (0)
Supported ongoing program, policy, practice	28 (62)	14 (31)	3 (7)	0 (0)
Led to initiation of new program, practice	23 (51)	19 (42)	3 (7)	0 (0)

Participants came to realize that constructivist principles guided the design of the academy. Based on Fosnot's (1996c) principles of constructivism relative to professional development, the format purposefully included activities that supported inquiry, encouraged questioning, and allowed time for the development of constructs; ideas that provoked disequilibrium balanced with the concomitant supports (i.e., collaboration and community) necessary to accommodate reestablishment of equilibrium; meaningful discourse and deep reflection; and recurring themes meant to convey big ideas. How these strategies influenced participants is articulated in some of the representative professional developer participant comments listed below.

> I valued the opportunities to discuss and share information, to reflect on our beliefs, and to learn as a constructivist rather than as a behaviorist. The variety of experiences and reflection that is required before understanding is achieved became clear; and I appreciated the value placed on my knowledge and experience. (Professional developer, academy final evaluation)

> As professional developers, we've got to understand the change process better and be able to assist educators through the process. The process takes longer and is more complicated than I thought. Learning to slow down challenged some of my traditional beliefs, caused me some discomfort, and stimulated a bit of internal tension. The spirit of collaboration within the academy community saw me through my struggle, and I've emerged a better change agent. (Professional developer, academy phase 1 evaluation)

> I was surprised to find that my ideas [about equity and who can learn] still need adjustment, even though I thought I had overcome all of my previous misconceptions. Now I'm awakened to the fact that I live in an area where equity does not seem to be a concern and equity issues are not being addressed. (Principal, academy final evaluation)

> I have begun to take notes on how facilitators craft their presentations. I have become interested in how the facilitator creates the moods and the flow—the transitions. And, I have proudly realized that I am a professional developer too. (Teacher leader, academy final evaluation)

Overall, participants' comments supported the value of inquiry with time to pursue it, disequilibrium with the community support to reestablish balance, discourse with time to reflect, and the construction of big ideas within the appropriate context needed to explore them. They reported that they valued the challenges presented by inquiry, disequilibrium, and discourse, and, as adult learners, found these challenges instrumental to the changes and improvements they were making in their own skills, habits, and attitudes.

Moving Through the Levels of Use

There is neither a generalized nor an ideal distribution across the CBAM Levels of Use continuum. The developers suggest, however, that a jump from nonuse to use requires support and information and that progression through the levels requires varying degrees of assistance. It is desirable, for instance, to concentrate efforts on moving individuals out of mechanical use whenever possible to prevent frustrations and inappropriate modifications to skills and practices.

On the basis of a number of data sources, including self-report data and observations from external sources, evaluators determined that none of the participants were in the first three levels of nonuse (Table 5.5, p. 74). Twelve (27%) of the participants demonstrated minimal incorporation of new skills and knowledge as a result of their academy experiences. They either used the new resources mechanically with awkward attempts at replication of activities ($n = 5$, 11%) or routinely with skilled replication ($n = 7$, 16%). These participants made little or no attempt to modify the materials from the academy or create new ones, and their concerns focused on their skills as presenters rather than on audience responses. However, 33 (73%) of the participants functioned more independently, indicated an awareness of their potential impact on others, and demonstrated behaviors at the three highest levels of the use continuum. Those at the refinement level ($n = 10$, 22%) reported modifying existing materials for use with their students or clients, those at the integration level ($n = 10$, 22%) reported actively encouraging others to try new ideas, and those at the renewal level ($n = 13$, 29%) reported using their skills to develop new materials and explore new practices.

Table 5.5 Levels of Use of new skills and practices of academy participants

Level	Total	Professional developer	Curriculum coordinator	Teacher	Principal	Super-intendent	Program director
n		21	4	12	3	2	3
3-Mechanical	5	1		3			1
4a-Routine	7	2		4			1
4b-Refinement	10	7		1		1	1
5-Integration	10	4		3	2	1	
6-Renewal	13	7	4	1	1		

Opportunity to practice, reflect, and enjoy serious discourse about practices and attitudes appeared to contribute to higher levels of use, and job responsibilities seemed to play a role in determining an individual's attained level of use. For example, of the 12 participants in mechanical or routine use, 7 were classroom teachers and only 1 of them had professional development responsibilities in her district. The 3 professional developers at these lower levels of use delivered predesigned workshops and had no input into the design. On the other hand, the three highest levels of use included the remaining 18 professional developers, all 5 district leaders (principals and superintendents), all 4 curriculum coordinators, and 5 classroom teachers who were department chairs with professional development responsibilities within their departments.

Reflections for the Future

The design of professional development is a complex endeavor. Although a variety of content process standards are available to guide the process, infusing a learning theory such as constructivism produces challenges. One of these is to continuously provide the right mix of activity and reflection, confrontation and collaboration, and content and process, and to allow for the unexpected. The reward is in watching learners struggle with a concept and emerge from the struggle with a new understanding. Inherent in the very nature of constructivism is the fact that there is no model, only guideposts for developers to follow as they create and design learning experiences for their students or adult learners.

Adhering to constructivist principles, creating a constructivist climate, and pursuing constructivist experiences for one's learners give pause to teachers (facilitators, professional developers) because the path is ever-changing and the end is not always in sight. The leaders are not often sure of where the learners' constructs will take them; hence, they sometimes question their own ability to make meaning in uncharted waters. Leaders must trust that their own disequilibrium is as important to the learning process as that of their learners. The academy facilitators designed activities and experiences for participants with no guaranteed or planned outcome, but they were skilled and confident in their own abilities to construct new knowledge from unanticipated results. This ability is critical to creating constructivist experiences for one's learners, whether they be school children or adult learners.

As for the academy participants, they all demonstrated a change in their use of skills and resources as defined by the CBAM Levels of Use. Their reported and observed new or improved practices included more attention to and understanding of the National Staff Development Council's staff development standards (NSDC 1996), greater focus on national science and mathematics content standards, increased emphasis on modeling constructivist techniques, incorporation of interactive and reflective strategies, and greater attention to the skills of facilitation. As opposed to the more traditional train-the-trainer models, the academy's developer-focused model nurtured in participants the ability to create their own resources as expected at the refinement, integration, and renewal levels of use, rather than merely replicating the ideas of others as defined by the mechanical and routine levels.

The academy model suggests that the skills and attitudes of professional developers regarding constructivist learning can be improved in a setting that models such practices and attitudes. The more STEM professional developers experience constructivist techniques in their own learning situations, the greater the chance that they will provide these types of experiences for their audiences. When one learns in a constructivist setting, there appears to be a greater chance of demonstrating new skills at higher levels of use. When capacity is built such that higher levels of use are internalized, the learners lessen their dependence on the script and the teacher or mentor and take charge of developing their own potential.

Conclusion

The lessons we learned and the strategies we perfected would seem to form a critical infrastructure for today's professional developers who are struggling to support other educators in their use of the *Common Core State Standards* (NGAC and CCSSO 2010) and the *Next Generation Science Standards* (NGSS Lead States 2013). We sincerely hope that they will find our work and findings supportive of their ongoing professional development enterprises and beneficial to their professional reflective thought processes.

References

American Association for the Advancement of Science (AAAS). 1989. *Science for all Americans*. New York: Oxford University Press.

Bringuier, J. -C. 1980. *Conversations with Jean Piaget*. Chicago: University of Chicago Press.

Bruner, J. S. 1971. *The relevance of education*. New York: Norton.

Champion, R. 1998. 'Sort of' is not getting us there. *Journal of Staff Development* 19 (4): 7.

Darling-Hammond, L., and M. W. McLaughlin. 1996. Policies that support professional development in an era of reform. In *Teaching learning: New policies, new practices*, ed. M. W. McLaughlin and I. Oberman, 202–218. New York: Teachers College Press.

Dewey, J. 1916. *Democracy and education*. New York: Macmillan.

Duke, D. L. 1990. Setting goals for professional development. *Educational Leadership* 47 (8): 71–75.

Duke, D. L. 1993. Removing the barriers to professional growth. *Phi Delta Kappan* 74 (9): 702–712.

Fosnot, C. T. 1996a. Preface. In *Constructivism: Theory, perspectives, and practice*, ed. C. T. Fosnot, ix–xi. New York: Teachers College Press.

Fosnot, C. T. 1996b. Constructivism: A psychological theory of learning. In *Constructivism: Theory, perspectives, and practice*, ed. C. T. Fosnot, 8–33. New York: Teachers College Press.

Fosnot, C. T. 1996c. Teachers construct constructivism: The center for constructivist teaching/teacher preparation project. In *Constructivism: Theory, perspectives, and practice*, ed. C. T. Fosnot, 205–216. New York: Teachers College Press.

Friere, P. 1970. *Pedagogy of the oppressed*. New York: Herder and Herder.

Friere, P., and I. Schor. 1987. *A pedagogy for liberation*. South Hadley, MA: Bergin and Garvey.

Greene, M. 1978. *Landscapes of learning*. New York: Teachers College Press.

Hall, G. E., and S. M. Hord. 1987. *Change in schools: Facilitating the process*. Albany, NY: State University of New York Press.

Heidegger, M. 1968. *What is called thinking?* New York: Harper and Row.

Holzer, S. M. 1994. From constructivism to active learning. *Innovator* 2: 4–5.

Julyan, C., and E. Duckworth. 1996. A constructivist perspective on teaching and learning science. In *Constructivism: Theory, perspectives, and practice*, ed. C. T. Fosnot, 55–72. New York: Teachers College Press.

Killion, J., and C. Harrison. 1997. The multiple roles of staff developers. *Journal of Staff Development* 18 (3): 34–44.

Knowles, M. 1978. *The adult learner: A neglected species*. Houston, TX: Gulf Publishing Company.

Martin, J. R. 1981. Sophie and Emile: A case study of sex bias in the history of educational thought. *Harvard Educational Review* 51 (3): 357–372.

McLaughlin, M. W., and J. E. Talbert. 1993. Introduction: New visions of teaching. In *Teaching for understanding: Challenges for policy and practice*, ed. D. Cohen, M. W. McLaughlin, and J. E. Talbert, 1–10. San Francisco: Jossey-Bass Publishers.

National Council of Teachers of Mathematics (NCTM). 1989. *Curriculum and evaluation standards for school mathematics*. Reston, VA: NCTM.

National Council of Teachers of Mathematics (NCTM). 1991. *Professional standards for teaching mathematics*. Reston, VA: NCTM.

National Governors Association Center for Best Practices and Council of Chief State School Officers (NGAC and CCSSO). 2010. *Common core state standards*. Washington, DC: NGAC and CCSSO.

National Research Council (NRC). 1996. *National science education standards*. Washington, DC: National Academies Press.

National Staff Development Council (NSDC). 1996. *Standards for staff development*. Oxford, OH: NSDC.

NGSS Lead States. 2013. *Next Generation Science Standards: For states, by states*. Washington, DC: National Academies Press. *www.nextgenscience.org/next-generation-science-standards.*

Piaget, J. 1973. *To understand is to invent: The future of education*. New York: Grossman.

Piaget, J. 1974. *The language and thought of the child*. New York: The New American Library.

Proust, M. 1934. *Remembrance of things past*. New York: Random House.

Regional Educational Laboratory Network. 1995. *Facilitating systemic change in science and mathematics education: A toolkit for professional developers*. Washington, DC: Regional Educational Laboratory Network.

Rowe, M. B. 1978. *Teaching science as continuous inquiry: A basic*. New York: McGraw-Hill Book Company.

Turnbull, B. J. 1995. True technical assistance views educators as learners. *R & D Preview: Traditions of Learning* 10 (5): 12.

Sparks, D., and S. Loucks-Horsley. 1989. Five models of staff development for teachers. *Journal of Staff Development* 10 (4): 40–57.

von Glasersfeld, E. 1996. Introduction: Aspects of constructivism. *In Constructivism: Theory, perspectives, and practice*, ed. C. T. Fosnot, 3–7. New York: Teachers College Press.

Chapter 6

Ohio's 30 Years of Mathematics and Science Education Reform: Practices, Politics, and Policies

Jane Butler Kahle and Sarah Beth Woodruff

Introduction

This chapter focuses on over three decades of science and mathematics reform efforts in Ohio. Beginning with mathematics projects—primarily developed at The Ohio State University (OSU) and funded by the Ohio Department of Education (ODE)—continuing with awards by the National Science Foundation (NSF), and culminating with changes in policy and increased funding of projects throughout the state by both the ODE and the Ohio Board of Regents (OBR), the two educational agencies in the state. The first part of this chapter describes the mathematics projects that laid the groundwork for Ohio's systemic effort. The main focus is Ohio's *Discovery* project, which was one of the ten initial Statewide Systemic Initiative (SSI) projects funded by NSF. The chapter goes on to discuss the transition projects that continued and expanded *Discovery*'s professional development.

The second part of the chapter describes changes at the state level and initiatives in mathematics and science at both state agencies. The primary objective of the second section is to describe how the reform was continued by both state agencies through Teacher Quality Improvement and Mathematics and Science Partnership awards in order to reach Ohio's rural as well as urban students.

We conclude each section by describing lessons learned, and we end by reviewing how coordinated, collaborative efforts that are sustained over 30 years can lead to quality mathematics and science education in a state, affect its political system, and contribute to changes in state educational policies.

Part I
Ohio's Systemic Reform

Rationale

Early Statewide Efforts in Mathematics

In the mid-1970s, mathematics educators and professors at over half a dozen Ohio colleges and universities began to think about changing the teaching and learning of K–12 mathematics. Together with the mathematics curriculum expert at the ODE, they began to offer professional development, often in collaboration with the Ohio Council of Teachers of Mathematics. The professional development initially focused on introducing and using the metric system; it continued with an emphasis on problem solving; and, later, it focused on the National Council of Teachers of Mathematics (NCTM) standards for K–12 mathematics (NCTM 1989). Although grounded in individual projects, the scope of the effort was impressive. As Steve Meiring, the mathematics supervisor at the Ohio Department of Education, recalls, 75 workshops were provided annually, reaching 2,200 teachers and disseminating $100,000 worth of curricular materials free to teachers from 1978 to 1991.

Meiring also recalls that the most significant effort in mathematics focused on changing the teaching and learning of mathematics from memorizing to problem solving. Problem-solving workshops were offered by teams of mathematicians and mathematics educators for teachers in five regions of the state in 1977, 1979, and 1981. After the NCTM standards embraced problem solving, hundreds of workshops were conducted by regional chapters of the Ohio Council of Teachers of Mathematics and university leaders who distributed over 30,000 of the ODE's problem-solving booklets. When Ohio developed statewide proficiency tests with a 50% emphasis on word problems, the reform became both systemic and sustained.

Meiring provides another example of how the mathematics reform set the stage for statewide systemic reform. During three summers, institutes addressing the organization and content of K–12 mathematics courses were offered for secondary mathematics department chairs. Each institute enrolled between 20 and 25 such chairs, essentially involving every area of the state.

Continuing with a systemic approach, a teacher in Columbus (Al Adcock) and a mathematician (Bert Waits) at OSU collaborated to develop and implement the Early Mathematics Placement Test (EMPT). The test was aimed at high school juniors whose coursework indicated their readiness for college mathematics in various majors. Based on individual student scores on the EMPT, recommendations were made for students' senior year mathematics courses. The point was to circumscribe, if not wipe out, the need for remedial mathematics in Ohio's colleges and universities. The program grew from seven high schools participating in 1978–1979, to 232 high schools in 1982–1983, to over

500 high schools by 1983–1984. Concomitantly, two professors at OSU (Frank Demana and Joan Leitzel) developed a 12th-grade high school course, Transition Mathematics, for low-performing EMPT scorers, which used problem-solving exercises to rebuild students' arithmetic skills into algebraic representations. Combined, these efforts were systemic in providing support for high school teachers and students and in addressing the historic problem for Ohio's colleges and universities of mathematics remediation.

In another important way, Ohio was moving toward systemic reform; that is, state proficiency tests (in both mathematics and science) as well as elementary and secondary content standards were developed. Further, the ODE developed model courses of study for mathematics and English language arts, and, initially, local district courses of study required approval by the ODE. These policies developed the idea of working up and down the educational system to change the practice of teaching and to improve student achievement. The mathematics reform involved ODE personnel, teachers, department chairs, and faculty. Further, aspects of the reform had won the support and approval of the Ohio General Assembly, and that recognition led to changes in policy. This 20-year effort to improve the practice of teaching mathematics laid the political foundation for the next stage of reform.

Model

Discovery: *A Systemic Approach to Reform*

Background

Ohio's reform effort was fortunate to have been funded by NSF and to have the support of both the Governor's Office and the OBR for the *Discovery* project. Project leadership was at two Ohio universities, OSU, with Kenneth Wilson as coprincipal investigator, and Miami University, with Jane Butler Kahle as principal investigator. The OBR served as the fiscal agent for *Discovery*. The combination of universities and government agencies guiding *Discovery* proved to be invaluable in obtaining state funds, in reaching large segments of the population, and, eventually, in sustaining the effort after the end of the initial funding period. With the support of two governors, NSF funds were matched each year by monies allocated by the Ohio General Assembly. Further, by the end of the federal cycle of funding, *Discovery* was a line item in the state budget (Kahle 2008).

Several early decisions, some of which addressed the practice of teaching science and mathematics and others of which focused on policies or politics in the state in the early 1990s, proved to be important. First, *Discovery* limited itself to middle school (loosely defined as grades 5–10). That decision was both practical and political because all students in those grades take mathematics and science. Second, as a populous state, the total combined funds could reach only some of its teachers, so the decision was made to focus funds where they were most needed. Because Ohio did not yet certify teachers specifically for middle school science or mathematics, many teachers of science and mathematics in grades 5–10 lacked sufficient coursework in those subjects. Therefore, middle school was the project's focus. Third, Reading Recovery provided a proven model for successful professional development. Although Reading Recovery used a trainer-of-trainers model, it had built-in safeguards to ensure fidelity of the professional development (Clay 1993). Fourth, *Discovery* adapted a model for developing and implementing new components from industry; that is, the research-design-redesign process used in designing aircraft. Fifth, it was decided that *Discovery* initially would target the 41 school districts that the ODE had categorized as "at risk."[1] These districts were primarily in Ohio's large cities or in isolated rural areas and enrolled high proportions of economically disadvantaged students.

Outreach and Awareness

During the proposal writing stage, various personnel at both state agencies insisted that a support network needed to be established for the project to be successful and sustained. The decision to divide the state into eight regional centers, each with a major media outlet, proved to be critically important—for organizational purposes as well as for outreach and public awareness. Each *Discovery* regional center had a director, a council, and, eventually, scientist educators and teacher leaders. Each center received its initial fiscal support from *Discovery*, and each sought additional funds from local, state, and federal sources as the reform progressed.

As the project developed, additional efforts to maintain public awareness of the importance of quality mathematics and science education included having *Discovery* teachers (participants in the professional development) contact their representatives to the General Assembly. In addition, the Miami University site conceived, developed, and distributed small, concise booklets, *Pocket Panoramas* (which fit into a pocket) that described, with graphs and tables, the impact of *Discovery* on Ohio's teachers and students of mathematics and science (Kahle and Rogg 1995, 1996, 1997; Kahle and Damnjanovic 1998; Kahle, Meece, and Damnjanovic 1999). These booklets were distributed annually to all state legislators, the governor, the chancellor of the OBR, the state superintendent of schools, Ohio's senators and congressional representatives, and the superintendents of Ohio's 614 public school districts. Last, short inquiry activities were developed for the state parks. These were distributed free as one entered a park, and many families were introduced to *Discovery* through them.

Improving Teaching Practice and Student Learning

In the early 1990s, one of the several groups evaluating the SSI program developed a model for the systemic reform of education (Zucker et al. 1998). The model contained five levels with states (or other educational agencies) generally using one of two approaches. First was the top-down approach. In that approach, initial efforts at reform occurred at the state level (usually from the governor or the department of education). Those efforts (e.g.,

---

1. The "at risk" category was composed of districts that had low scores on Ohio's initial competency tests.

additional funding of selected schools to create model schools or new content standards) then affected the other levels—districts, schools, teachers, and students. Efforts initiated at the local level of an educational system, primarily through professional development of science and mathematics teachers, were characterized as bottom-up reforms. The developers of the model suggested that reform could enter a system at any level, but to be successful it had to affect all levels of an educational system, in this case a state. *Discovery* used a dual approach, enlisting the support of the two educational agencies to address certain policies (e.g., middle school mathematics and science teacher licensure) in the top-down model while using the bottom-up approach with an emphasis on teacher professional development to change the practice of teaching and learning science and mathematics in schools and districts.

Discovery*'s Professional Development Program*

In addition to adapting a delivery model from Reading Recovery and a design model from industry, *Discovery* adopted a successful undergraduate curriculum for its inservice teacher professional development. Under the leadership of Arnold Arons and Lillian McDermott, the University of Washington had revolutionized the teaching of undergraduate physics to preservice elementary teachers (Arons 1990; McDermott, Shafer, and Rosenquist 1996). The model, *Physics by Inquiry*, involved inquiry into natural phenomena, cooperative learning groups, and open-ended questioning. It provided both the pedagogy and the content of the initial *Discovery* professional development institutes. However, the mode of delivery had to be determined. Based on research indicating that sustained changes in teaching required over 120 hours of professional development (Loucks-Horsley et al. 1990), the project implemented six-week summer institutes, followed by four to six half-day seminars during the school year. In its first summer, *Discovery* institutes were provided at OSU and at Miami University for teachers in the regional centers surrounding the media outlets of Columbus and Cincinnati. *Physics by Inquiry* provided the curriculum, and the institutes were taught by university faculty who had attended seminars at the University of Washington and who had taught—or were planning to teach—*Physics by Inquiry* to undergraduates.

At the same time, mathematicians and biologists at the two host institutions were identified to develop and teach institutes in mathematics and life science. All three content-based institutes eventually were provided for teachers at 21 colleges and universities across Ohio. And, as the project progressed, the institutes were taught by a combination of faculty, scientist educators, and teacher leaders, all of whom initially participated in an institute alongside middle school teachers. Doctoral-level scientists in physics and biology as well as in mathematics were recruited and selected in each region to become scientist educators. They were released full-time from their previous positions and were paid by *Discovery*. Likewise, teacher leaders in physics, life science, and mathematics were recruited and selected in each region. At the teachers' request, they were released from their teaching duties one-half time, with *Discovery* paying one-half of their salaries. These skilled and highly qualified professionals not only team-taught the summer institutes and school year seminars but also provided in-school support for *Discovery* teachers who were changing teaching and assessment practices in their classrooms.

Although administrators, usually a *Discovery* teacher's principal, were invited to attend a day at the summer institutes, the project did not focus on them. It was clear that most administrators were uncomfortable at the institute sessions, probably due to a lack of confidence in their science and mathematics knowledge. Although the one-day sessions were continued, the project did not increase its outreach to principals. This lack of attention was unfortunate, because some administrators failed to understand the "hows" and "whys" of teaching through inquiry. One poignant example was an annual teaching evaluation, shared by a *Discovery* teacher. She was evaluated while teaching a unit on the solar system. Her students had measured and converted various distances and were positioned about the room and hall representing the sun and its planets. Movements of the planets were demonstrated when the students moved. She received a rave evaluation from her principal on several items; however, he concluded by adding a note that he would buy her a model of the solar system to use next year.

Evaluations of Discovery

Discovery was fortunate that Iris Weiss, Horizon Research Institute, agreed to be its evaluator. Horizon not only provided deep insights into what was working and why; it provided technical assistance to help change or modify direction. The expertise of the Horizon staff was of enormous value throughout the project.

In addition, midway through the systemic effort, *Discovery* mounted its own internal evaluation. The principal investigators realized that data were needed in order to convince state leaders to continue to support the project after federal funding. The Miami University site developed valid and reliable questionnaires about teaching and learning practices; and, in collaboration with faculty, scientist educators, and teacher leaders across the state, it developed two *Discovery* inquiry tests—one in mathematics and one in science. Based on a three-tier research design that involved both quantitative and qualitative techniques, data were gathered and analyzed annually, and the findings became the backbone of the *Pocket Panoramas* (Kahle and Rogg 1995, 1996, 1997; Kahle and Damnjanovic 1998; Kahle, Meece, and Damnjanovic 1999).

Redesigns and Scaling-Up of the Model

Although the original plan of sustained, content-based institutes taught by inquiry did not change, many components of the original model evolved to meet new demands and needs. For example, the initial plan was to bring two regions to capacity each year; however, popular demand necessitated a change and all regions were fully participating, offering institutes in physics, mathematics, and life science, by year 3, with the state allocating monies to support the expansion. Lobbyists at Ohio's universities and colleges as well as at the state educational agencies were key in this effort, but *Discovery* teachers who contacted their own legislators did the most effective political outreach.

A few years into the project, *Discovery* leaders realized that in spite of active recruitment, minorities were underrepresented as scientist educators and as teacher leaders. *Discovery*, then, offered to support an additional position in each region, if the position were filled by a scientist or teacher belonging to an underrepresented minority in science or mathematics. Over half of the regions identified outstanding minority candidates, all of whom contributed greatly to the project.

Although equity was a theme of Ohio's SSI from its inception, equity tended to get lost after participating teachers returned to their classrooms. They had their plates full teaching in a new way and, perhaps, convincing administrators and parents that memorizing was not the best method by which to learn science or mathematics. Therefore, three outstanding teacher participants from different parts of the state were asked to form an equity team. Building on their own experiences, they developed programs, pamphlets, and lessons focused on equity; then they traversed the state, helping to infuse equity into all aspects of the project.

As indicated above, *Discovery* entered the pyramid model for systemic reform at both the top and bottom levels. Which level was emphasized varied as the project matured. Clearly, in the beginning, the emphasis was at the district or school level as teachers were recruited and selected from the 41 underachieving districts. In large urban districts, *Discovery* focused on identifying teams of teachers in the same school, whereas in isolated rural districts it aimed to attract teachers in several schools who were teaching the same grade or subject. The point was to build a critical mass of *Discovery* teachers in a school or district who could (and would) influence changes in teaching and learning. After student achievement on Ohio's proficiency tests in mathematics and science and on *Discovery*'s inquiry tests[2] indicated that having more than one *Discovery* teacher in a school contributed significantly to enhanced test scores, the project actively recruited teachers in the same school. When Ohio began using the state tests to rate school effectiveness, *Discovery* directed its teacher recruitment to the schools and districts with the lowest performing students.

With the regional centers in charge of teacher recruitment and institutes, *Discovery*'s statewide leadership (personnel at both state educational agencies, the principal investigators, and a coordinating council with representatives from industry, professional associations, academia, and K–12) directed more of its activities to the top levels of the pyramid. As mentioned, extensive efforts were made to enlist the support of legislators and superintendents across the state. In addition, *Discovery*'s scientist educators, teacher leaders, and institute participants were instrumental in contributing to the development of new state standards, in recommending new science and mathematics courses and curricula, and in conceptualizing a new licensure system for teachers of mathematics and science. Those

2. The *Discovery* mathematics test correlated with the state test at $r = 0.97$, while the science test's correlation with the state test was $r = 0.98$.

changes, spearheaded by *Discovery* teachers, have institutionalized and sustained not only the spirit of the reform but also a deep understanding of what is required to change a system in order to enhance teaching and learning. For example, principals' and teachers' responses to a questionnaire concerning what constitutes effective professional development showed a change from short school day activities to sustained activities outside of the school day over the period of *Discovery* (Kahle, Meece, and Damnjanovic 1999).

Bridging the Gap: *Equity in Systemic Reform*

An important follow-up project was awarded to *Discovery*'s leadership at Miami University (Kahle, Rogg, and Tobin 1996). The project, built directly on *Discovery*'s evaluation activities, continued to collect data about student learning and teaching practices and—most importantly—used sophisticated statistical tests to analyze the data. Bridging the Gap: Equity in Systemic Reform was designed specifically to assess any narrowing of identifiable achievement gaps between boys and girls, between African American (Ohio's largest minority group) and white students, or between students from different economic levels. It was guided by three research questions and involved levels of data collection, levels A, B, and C.

In 1994, 150 schools were randomly selected to participate in *Discovery*'s internal evaluation, and from 1995 to 1999, over 100 of those schools participated in Bridging the Gap. At those schools, designated in the study as level A, principals and all mathematics and science teachers (grades 6–9) completed questionnaires focusing on standards-based teaching, student learning, parental involvement, and administrative support for science and mathematics education.

Level B consisted of a subset of the original random sample of schools. Across the years, the number of schools agreeing to participate in level B ranged from 12 to 16. The following criteria were used to select level B schools: (1) They were part of the statewide random sample (level A); (2) they enrolled approximately 30% African American students; (3) they had at least one teacher who had participated in *Discovery*'s professional development programs; and (4) high proportions of their students were eligible to receive free or reduced-price lunch. At level B schools, students completed a questionnaire that included items parallel to items on the teacher and principal questionnaires, and they also completed either a science or a mathematics achievement test designed to measure their problem solving abilities and conceptual understandings. In addition, for four years (1994–1997) three- to four-day site visits were conducted at level B schools.

Level C involved intensive two- to three-year case studies in five schools. Extensive school and classroom observations were conducted. In addition, students, teachers, administrators, parents, and community leaders were interviewed (Kahle, Rogg, and Tobin 1996).

Some of the most important findings of the Bridging the Gap study involved the frequency of use of inquiry and problem solving (standards-based) teaching practices and enhanced student learning. A carefully controlled study focused on student achievement in eight level B urban junior high and middle schools (Kahle, Meece, and Scantlebury 2000).

Frequency of use of reform teaching practices was reported by both students and teachers, with student responses validating those of their teachers. Student and teacher data were linked, and teachers were identified as having participated (or not) in *Discovery*'s professional development. The same measure of achievement was used in all of the schools. In the end, results were analyzed only for African American students, because the number of white students in the schools was too low for statistical analyses. Using hierarchical linear models analyses, the researchers found that

> students of teachers who participated in the SSI professional development, compared with students whose teachers had not participated, scored higher on the science achievement test. Second, students of SSI teachers rated their teachers as more frequently using standards-based teaching practices than did students in non-SSI teachers' classes. (Kahle, Meece, and Scantlebury 2000, p.1034)

However, participation in professional development provided by *Discovery* was not a significant predictor of student achievement, rather the use of standards-based teaching practices was predictive. The acceptance and use of inquiry and problem solving may be the longest lasting impact of the reform.

In addition to NSF funding for Bridging the Gap, Ohio's General Assembly provided funds for *Discovery* at Miami University and for Schools and Universities Statewide Teaching Approaches to Inquiry (SUSTAIN), at OSU from 1997 to 2001. Through SUSTAIN, collaborations were formed between Ohio's largest school districts and institutions of higher education. Its focus was on improving undergraduate education in mathematics and science, particularly for prospective teachers. The collaborations included eight urban districts that employed over 14,000 teachers as well as the universities educating the largest numbers of prospective teachers. The continued efforts of *Discovery* with state funding are described next.

Building on Discovery

As the first phase of funding expired, Ohio committed to providing funds for the continuation and expansion of mathematics and science education systemic reform efforts. In 1997, Ohio's General Assembly initially funded *Discovery* for two years with mandates to expand the initiative to elementary and high schools, to improve the content knowledge of all teachers of science and mathematics, and to continue to study and report on the effectiveness of Ohio's reforms. Between 1997 and 2007, the OBR funded mathematics and science education reform, under the *Discovery* name, in excess of $6 million. These funds were used to provide mathematics and science programs primarily for teachers but also for school and district administrators. Through professional development institutes of one to six weeks, including a two-week experience focused on whole-school reform, *Discovery* continued to provide sustained, content-rich, standards-based, inquiry-oriented experiences for Ohio educators.

In 2002, the OBR awarded funds to Miami University to develop an online teacher professional development experience (*iDiscovery*) that would follow up various face-to-face science and mathematics professional development programs. The *iDiscovery* project supported sustained professional development by establishing and facilitating online learning communities for science and mathematics educators across the state. Since its inception, *iDiscovery* has received more than 2.9 million dollars in state funding and has served more than 13,000 teachers, including those in isolated rural schools without ready access to opportunities for professional learning.

Lessons Learned

Implications for the Next Phase of Reform

The following list of lessons learned from *Discovery*, Bridging the Gap, and various follow-up projects is arranged in chronological order, not in order of importance. Some are small lessons and others are major ones that, if ignored, would have either ended or diminished the reform. All resulted in adjustments that eventually led to scaling up and sustaining the reform for over 30 years.

- Systemic change requires a threefold process: first, the vision must be articulated and broadly communicated; second, the infrastructure must be changed and strengthened; and third, the infrastructure must embrace and support the reform.
- Systemic reform is an evolutionary process with each change necessitating other ones.
- A regional delivery and support system is necessary in a large, populous state.
- There is neither time nor resources to do all things. In Ohio's case, the reform used existing, research-based curriculum rather than attempting to develop new materials.
- One study of the SSI projects concluded that centrally managed projects, such as Ohio's, that featured intensive teacher training were the most successful (Freundlich 1998).
- Equity still requires affirmative action at all levels of a project or a reform movement. Leaders cannot assume that others understand equity beyond the notion of equal numbers of girls and boys or appropriate proportions of majority and minority students studying mathematics and science.
- Charting the progress of the reform and publishing the findings, including any changes in student learning, helps build a foundation of trust and collaboration and sustain the reform through additional funding.
- All parts of a reform must be addressed and work together if the results are to be systemic. As a reform progresses, alterations must lead to wider participation and acceptance of the reform.
- Research or assessment without dissemination benefits only those who are already involved in the reform. Dissemination of findings

in practical and easy-to-use ways informs others of the success and invites them to become active participants or supporters.

- The pool of teachers who volunteer to undertake sustained professional development is limited. Nonvolunteer teachers must be reached in their communities and schools, and the professional development must be at the level at which they teach mathematics or science. *Discovery* evolved to meet this challenge by offering a series of 40-hour workshops taught by *Discovery* teachers and using grade-appropriate, research-based curricula. Districts supported the teacher-instructors and often required all science or mathematics teachers at the targeted grades to attend.
- Teachers, principals, and parents are reluctant to agree to experimental studies; rather, comparison groups of students and teachers may have to be used. Therefore, patterns in the findings became important. For example, student achievement was assessed by comparing scores of students of *Discovery* teachers with those of students whose teachers had not participated on *Discovery* tests *and* on State Proficiency Tests in mathematics and science.
- For changes in K–12 education to be sustained over time, teacher preparation must be addressed in the reform.
- Electronic networks are effective and efficient ways to provide support for teachers as they endeavor to implement new instructional and assessment strategies.

Overall, *Discovery* and its offshoots were successful beyond expectations. However, as *Discovery*'s leaders moved on and as its teachers gradually retired, other initiatives—still focused on quality teaching and learning in mathematics and science—continued the reform. Those initiatives, largely supported by federal funds to the two state educational agencies, are discussed next.

Part II
Ohio Mathematics and Science Partnership Program, Sustaining Reform

Rationale

Teacher Quality Improvement and the Mathematics and Science Program

As discussed above, after NSF funding ended, the state continued to support *Discovery*. In 2002, following the passing of the No Child Left Behind Act, Ohio—like all states—received new sources of federal monies for science and mathematics education. The Improving Teacher Quality (ITQ) part of the funds was administered by the OBR, while the ODE handled the Mathematics and Science Partnership (MSP) funds. Together, these federal funds allowed the state to continue its reform of science and mathematics education on a large scale.

Background

Sweeping policy changes followed on the heels of the *Discovery* initiative. Changes to Ohio's teacher licensure system, the development of state academic content standards (ACS) in science, the implementation of a new state assessment system as well as more rigorous high school graduation requirements consumed the attention of policy makers and education stakeholders through the initial part of the new decade. These changes, many of which can be attributed to *Discovery*'s influence, were aimed at fortifying mathematics and science education in Ohio and addressing system issues that might impede gains in student achievement. In order to sustain and advance the reform initiated by *Discovery*, policy changes were also necessary.

Ohio addressed the need for a different type of certification for teachers of middle school and overhauled the state teacher licensure system in 1998, with the most significant impact on science teacher credentials. Added to the teacher licensure system was a middle childhood license (grades 4–9) that required more content preparation than the typical elementary teacher credential (grades K–8). Further, licensure requirements for teachers of high school science were strengthened by aligning teaching credentials with ACS and with the disciplines of science, rather than with specific higher education courses. Simultaneously, the state revised and implemented new ACS in mathematics in 2001 and in science in 2002. These new standards provided the frameworks for the development of new assessments that moved beyond recall and that incorporated levels of cognitive demand, requiring all students to demonstrate their skills and understandings in a variety of ways. The revised Ohio science ACS explicitly described expectations for student proficiency in scientific inquiry, science and technology, and scientific ways of knowing. Similarly, the Ohio mathematics ACS described expectations that students would learn mathematical processes, including problem solving, reasoning, communication, representation, and connections. Items on the new state assessments evaluated student understanding of both content and process. Finally, in 2005 the General Assembly adopted new graduation requirements for students, the Ohio Core, which specified three required units of science and increased the mathematics requirement to four units of study.

At the federal level, the No Child Left Behind Act of 2001 (Public Law 107-110) became law in January of 2002. Monies were allocated to states through ITQ. The purpose of the ITQ program was to increase the academic achievement of all students by helping states and schools improve teacher quality. Through the ITQ program, state and local educational agencies received funds on a formula basis. Funds received by the ODE were designated as MSP monies,[3] whereas those received by the OBR were ITQ funds. Both state agencies offered competitive grants to colleges and universities to form partnerships between their schools or colleges of education and arts and sciences and with high-need schools. Both programs supported sustained and intensive high-quality science and mathematics professional development in order to ensure that teachers were able to improve instruction and enhance learning.

Although they were similar in many ways, fundamental differences existed in the MSP and ITQ programs in Ohio. The OBR awarded approximately $3 million in ITQ federal funds annually, whereas the MSP program, administered by the ODE, awarded approximately $5 to $6 million annually. For this reason, MSP projects were larger in scale and scope. ITQ projects were typically funded for 1 year, although those demonstrating effectiveness could receive additional funding. In a typical year, the OBR awarded between 20 and 30 grants for mathematics and science projects. Projects generally were small scale, localized, and without sufficient resources for sustainability, scalability, or even rigorous evaluations. Although ITQ projects were required to conduct local evaluations and to report on impact and outcomes, a program-level evaluation has not been performed.

MSP projects could be awarded for a maximum of three years; therefore, they were able to focus on sustained professional development efforts. The ODE made fewer MSP awards, historically not more than 10 in any year, and provided a high level of project oversight and support. In four funding cycles, a total of 23 MSP projects were funded. Further, while federal law required that ITQ grants be equitably distributed by geographic area within a state, MSP had no such mandate. In addition to the federally required local evaluation of each MSP project, Ohio's MSP program has been externally evaluated since 2005. For this reason, more is known about how the MSP projects have impacted mathematics and science education in Ohio. MSP will, therefore, be the focus of the remainder of this chapter.

Improving Teacher Professional Development

In the early 2000s, the implementation of new mathematics and science ACS created an immediate need for developing teachers' awareness of standards-based teaching and of its implications for classroom practice and student learning. Mathematics education was funded by a substantial line item in the state budget; therefore, ODE mathematics

---

3. Between 2002 and 2003, the U.S. Department of Education MSP funding jumped from $12.5 million to $100 million.

education leaders were able to provide support for teachers immediately after the release of new mathematics ACS. Science did not have access to sufficient state funds, so the science reform was supported by MSP funds.

Ohio's first effort (2002–2005) at implementation of professional development programs, using federal MSP funds, resulted in four statewide projects designed to meet teachers' needs related to standards-based mathematics and science education and to reach large numbers of teachers in all regions of the state. Both the mathematics (funded through Ohio's Mathematics Initiative) and the science projects (supported by MSP funds) were for teachers of grades 7–10. Professional development projects for mathematics and science teachers in grades 3–6 were developed and piloted in 2005, with both subjects supported by MSP funds.

The professional development materials and curricula, including content-based modules and activities, support materials, and facilitation guides, were developed by science, mathematics, and education faculty in collaboration with teams of master teachers, curriculum specialists, and ODE content experts. Funding for materials development was awarded through a competitive grant process and exceeded $1 million for each project. The projects were managed through Ohio's existing regional professional development system[4] in an attempt to establish a systemic and uniform delivery mechanism based on the *Discovery* model. Also following the lessons learned in *Discovery*, the projects used a trainer-of-trainers model to prepare facilitators who led regional professional development sessions. Each project consisted of an intense, content-focused, one-week (40 hours) summer institute and a few days of follow-up during the subsequent school year. Teachers were expected to participate in the project for only one summer and academic year. Between 2002 and 2005, 12,000 teachers of science and mathematics participated in these professional development projects across all 12 regions of the state.

Although these initial MSP projects met federal requirements, ODE content experts realized the need to move beyond an awareness model in order to develop the capacity to improve science and mathematics education through school- or districtwide reform. In 2005–2006, assessment of the four MSP projects suggested the need to revisit the Ohio model for teacher professional development. The evaluation focused on fidelity of professional development implementation, which was found to be very low, and impact on teacher content knowledge learning, which varied across programs and regions. Findings suggested that variability in implementation might account for differences in teachers' experiences and outcomes (Feder, Horwood, and Woodruff 2006).

The ODE's change in approach represented a redefinition of systemic reform that placed less emphasis on the state system and greater emphasis on local districts as the target for reform. State education leaders used lessons learned from *Discovery* and the initial MSP projects to develop a solicitation that (1) focused at the local or regional level, (2) defined teacher and student needs, and (3) drew upon strengths and expertise of project

4. The regional professional development system in Ohio has gone through several iterations having different names; these agencies are currently called State Support Teams.

partners. More stringent criteria for projects were developed and communicated in a new MSP solicitation in December 2005. The ODE Office of Curriculum and Instruction was (and remains) responsible for the administration of the MSP program in Ohio. Education leaders in this office outlined requirements for projects to (1) develop *research-based* professional development models, (2) deliver professional development to *inservice* teachers, (3) provide *sustained* professional development (120 hours over two years), (4) include district and building *leadership* in the training, (5) use high-quality professional development materials that *meet school partners' needs,* and (6) reach a *critical mass* of educators by actively recruiting and supporting teams of teachers from a school or a district.

Reflecting on *Discovery* and findings of the follow-up Bridging the Gap study, ODE professionals also sought to address equity in the revised solicitation by requiring partnerships with high-need schools and implementing strategies to improve the academic achievement of traditionally underserved students. These recommendations were based on research that demonstrated the impact of inquiry-based instruction on the academic achievement of underserved students in urban and rural districts (Rosebery, Warren, and Conant 1989), on students' scientific literacy (Lindberg 1990), and on critical-thinking skills (Narode et al. 1987). The new Ohio science ACS specified that students were expected to (1) use inquiry processes to ask questions; (2) gather and analyze information; (3) make inferences and predictions; and (4) create, modify, or discard some explanations (ODE 2003). Mathematics content standards indicated that students were expected to (1) apply problem-solving and decision-making techniques and (2) communicate mathematical ideas (ODE 2002). The 2005 Request for Proposals (RFP) explicitly required projects to focus on inquiry as a vehicle for improving student achievement. In making this requirement, the ODE also outlined how projects should provide inquiry experiences for teacher participants.

In addition to focusing on developing deeper content knowledge, the revised MSP solicitation laid the foundation for the Ohio MSP Program to have systemic impact on mathematics and science teaching and learning. Several of the principles, outlined in the RFP, went beyond requirements for projects; they were pronouncements of the ODE's expectations for Ohio's teachers and their students. For example, specific references were made to (1) examining the developmental progression of student content knowledge and expectations across grade levels; (2) helping teachers understand student thinking, developing strategies to reflect on what students know, and finding effective ways to address common misconceptions; (3) promoting appropriate, authentic, and effective uses of classroom technology; and (4) developing teachers' abilities to use content-specific pedagogy and a variety of assessment strategies. In addition, the new RFP emphasized the joint responsibility of Ohio's K–12 educators and higher education faculty to improve mathematics and science teaching and learning by clearly laying out a model that tapped the strengths of both groups.

Model

Ohio's MSP Professional Development Program

Between 2006 and 2010[5], 23 MSP projects were funded, each with an award of up to $550,000 for the first year and with larger awards possible in subsequent years. All Ohio MSP projects addressed either mathematics or science or both and focused on elementary or secondary grade levels or both. Projects were designed and implemented to meet the unique needs of each partnership, so they varied widely in content and structure as well as in the experiences provided for teachers. Successful projects were eligible for two additional years of funding. Projects awarded in the first round of funding (2006) tended to be more comprehensive, developing programming for multiple grade levels or content areas. Later, the ODE clarified expectations and suggested that projects limit their focus to fewer grade levels and content areas, eventually requiring that partnerships initially concentrate on a single target school, a few schools, or one district and to either mathematics or science.

Beginning in 2005, the ODE funded four state-level evaluations of Ohio's MSP program. Ohio's commitment to evaluation made it possible to more closely align the MSP program with research on mathematics and science teacher professional development while simultaneously meeting state and local needs. Statewide program evaluations synthesized project-level findings to allow cross-project comparisons concerning the strengths and limitations of individual projects as well as to provide an evaluation of the Ohio MSP Program. Those findings, regarding the effectiveness of individual projects and of the MSP Program, informed funding decisions.

Phase 1: Realizing Federal and State Objectives for the MSP Program

In March 2006, nine Ohio MSP partnerships were funded, involving 29 Ohio colleges and universities and over 147 school districts. Each project was designed to meet the unique needs of its partner school districts. In March 2007, six additional projects were funded; those projects were required to address a growing body of research on professional development of mathematics and science teachers and the lessons learned from previous Ohio MSP projects.

In addition, a state-level evaluation was funded. The focus of the 2007–2010 MSP Program evaluation was on a policy-level synthesis of programmatic strengths and limitations using data from local evaluations, interviews of project personnel and participants, and document review. Components of the evaluation included a cluster analysis (Woodruff, Hung, and Seabrook 2009), an instrument study, and an effect size study (Woodruff, Li, and Kao 2010). These components resulted in important findings regarding characteristics of the emerging professional development models, and reliable as well as valid cross-project instruments. Further, it opened up the possibility of measuring impact on participants through meta-analyses.

The evaluation collected data from 15 Ohio MSP projects (funded in 2006 and 2007), which included more than 2,000 teacher participants. The projects varied widely in content

5. In 2005, six pilot projects in mathematics were funded with a combination of MSP and state Mathematics Initiative funds under a non-MSP solicitation.

and context, with six projects addressing science, five focusing on mathematics, and four involving both content areas. Elementary teachers made up 65% of all participants, while high school teachers represented only 10% of all participants. One hundred forty-seven partnering schools, including large urban, suburban, and very small rural schools from all regions of the state, were involved; 85% of the schools were identified as high-need.[6] Between 2007 and 2010, over 60,000 Ohio students were instructed by teachers participating in the MSP projects.

All 15 Ohio MSP projects, funded in 2006 and 2007, claimed to have made measurable and valid improvements in teachers' content knowledge (using locally developed assessments). Similarly, all projects reported having achieved measurable improvements in pedagogy, but for most projects this claim was not substantiated by classroom observation or by triangulation with other data. Four of the 15 projects reported measurable improvements in student performance on state assessments, with 2 reporting that the changes were statistically significant, but none reported student data for a comparison or control group (Zorn et al. 2009). Unfortunately, given the limitations of available data, a direct cross-project assessment of teacher or student impact was not feasible. In an attempt to report some degree of programmatic and project-level impact, the evaluators turned their attention to data from an instrument used by all projects to measure change in teachers' instructional practices. All 15 projects used the same teacher (self-report) perception instrument[7] to measure impact on instructional practices. A rigorous instrument study was undertaken, and findings suggested that all of the instrument subscales were reliable (Cronbach α 0.69 to 0.94) when used by each project and that all instrument subscales demonstrated good to very robust construct validity.

An effect size study, using data collected by the teacher instrument, found that the impact of the Ohio MSP professional development on self-reported teacher practices varied across projects and across program aspects. The largest programmatic effect sizes were found regarding teachers' preparedness to engage in standards-based instruction (pooled effect size [d] = 0.94) and their expertise regarding the use of standards-based instruction (d = 1.07). The MSP program had a moderate impact on changing teachers' reported instructional practice and a small impact on teachers' perceptions of the importance of using standards-based instruction. Not surprisingly, some projects affected participating teachers more significantly than did others (overall effect sizes [d] ranged from 0.02 to 1.77). All projects were at least moderately successful at helping teachers feel as though

6. A high-need school was defined as one that (1) has student achievement scores below 75% on statewide assessments for mathematics or science, or (2) serves high numbers of students from families with incomes below the poverty line, or (3) employs a high number of teachers who do not have "highly qualified" status in the academic area or subjects they are assigned to teach.

7. The teacher instrument was developed by Caliber and Associates and ODE personnel for use in the original MSP statewide projects in 2004. This instrument was subsequently modified by new MSP projects and used to measure four aspects of teacher practice: instructional practice, importance of use of standards for instruction, preparedness for use of standards for instruction, and expertise level of use of standards for instruction.

they had gained expertise in using standards-based instruction, yet no project significantly changed teacher attitudes about the importance of standards-based instruction (Woodruff et al. 2010).

Although the MSP projects were not assessed for quality of their professional development by local evaluations or by the statewide evaluation, the cluster analysis performed as part of the 2007–2010 statewide evaluation provided insights into how aspects of the professional development, specifically ones that research indicated could change teaching behaviors and enhance student learning (Desimone et al. 2002; Guskey and Yoon 2009), were addressed by the projects. An analysis of project professional development according to those features informed ongoing MSP work as well as the 2010–2012 MSP cross-project evaluation.

Phase 2: Scaling Back to Scale Up

In 2008, ODE again revised its strategy regarding the MSP Program.[8] In early 2008, several potential grantees were provided small planning grants to develop partnerships between high-need schools/districts and college and university faculty. In order to receive MSP funding, each partnership was required to conduct a needs assessment, articulate a tailored professional development plan of at least 120 hours, and ensure the engagement of all partners. It was ODE's expectation that school partners, rather than higher education partners, would take the lead on the new projects. Following the planning phase, two MSP science projects received funding. Both focused on whole-school reform with teacher professional development embedded in the school year. Although the projects only directly involved two schools, they piloted models of teacher professional development that were highly collaborative, more responsive to teacher and student needs, and systemic at the school level.

In 2008, the third evaluation of the Ohio's MSP program assessed the two science projects described above. Because the projects were fundamentally different from earlier ones, their evaluation was separate, but parallel, to the state level evaluation. It spanned both the planning and implementation phases of the two projects. It focused on the collection of primarily qualitative data, including observational classroom data, and its findings informed future project and evaluation efforts. An important finding was that the expectation that schools could provide the leadership necessary for the projects to succeed was not met.

Phase 3: Improving Capacity for Systemic Reform

In January 2010, a fourth round of MSP projects was funded. The solicitation contained all requirements of the 2008 solicitation and reflected the thinking of ODE personnel regarding sustainability and capacity building as mechanisms for furthering systemic mathematics and science education reform. ODE emphasized that projects should adopt or modify professional development models that were based on research and could be transported to other schools. Priorities for new MSP-funded projects were to (1) meet the specific needs of target school partners, (2) embed professional development in the academic year, (3) engage

8. Also in 2008, a separate MSP solicitation was issued exclusively to fund coaching projects in mathematics. One project was funded; it was administered and evaluated independently from others funded in that year.

all teachers in the target schools, (4) produce measurable changes in classroom practice, and (5) conduct a rigorous local evaluation of the project's progress and outcomes. Five projects were funded, three of which focused on mathematics and two of which focused on science. All projects targeted a single school site, or small district, and engaged all teachers of either mathematics or science in their professional development activities.

Concurrently, the ODE issued a solicitation for a state-level evaluation. Both the project and evaluation solicitations focused on capacity building as a mechanism for advancing systemic reform. Though MSP projects had largely been successful in delivering content-rich professional development to teachers, none had met expectations regarding the implementation of sustainable and scalable models that could be replicated. As a result, projects did not focus on developing research-based models. Further, projects had been only moderately successful at evaluating their outcomes and at making system changes to support and accelerate teacher improvement beyond the life of the project. Generally, projects had not focused on the quality of their professional development initiatives or on linking particular aspects of their professional development programs to teacher and student outcomes.

Drawing on the processes and findings of all prior state-level evaluations, the 2010 Ohio MSP cross-project evaluation focused on aspects of projects that research and experience indicate are critical to improving teacher and student outcomes. Data were collected and analyzed with attention to (1) effectiveness of projects' professional development implementation, (2) teacher participation and engagement in professional development, (3) professional development quality, (4) implementation of local evaluation plans, (5) partnership characteristics, (6) collaboration, and (7) attention to sustainability. Studying these aspects of projects, individually and collectively, provided a lens through which to view systemic reform at local and state levels. The evaluation team was composed of members of all previous state-level MSP evaluations and used lessons learned not only to provide a substantive cross-project evaluation but also to improve the rigor of local project evaluations.

The cross-project evaluation also implemented a system of technical support for local project evaluators in order to develop capacity through cross-project activities in collaboration with the state-level evaluators. Although the projects varied in content, approaches to project development and content delivery were more uniform than in the past, which permitted a more consistent application of review criteria and the use of common standardized instruments and data collection protocols. Further, the cross-project evaluation addressed the larger policy issues concerning successful implementation of the MSP program.

Lessons Learned

Lessons learned from the Ohio MSP program are similar to and different from those learned from *Discovery*. Because *Discovery* laid the foundation for systemic reform of mathematics and science education at the state level, a number of policies were in place to support the MSP initiative. An aligned system of ACS, state assessments, and clear expectations for teacher preparation and student achievement facilitated the MSP program primarily at the state level. For these reasons, the MSP Program's approach to systemic reform was focused largely at the local level.

- The value of authentic and engaged partnerships cannot be overestimated when system change is the goal. MSP's focus on partnership beginning with projects funded in 2006 has resulted not only in high-quality professional development experiences for teachers but also in an improved capacity of all partners to advance common reform goals.
- Synergistic activities of MSP partnerships have resulted in informal networks of mathematics and science educators, mathematicians, scientists, state agency personnel, and industry representatives who share similar perspectives regarding education reform that they communicate to political leaders.
- Federal, state, and local goals regarding mathematics and science education are similar, but competing priorities must be addressed. For example, persistent tensions among the value preferences of local project staff regarding evaluation sometimes conflict with those of funding agencies that set strict standards for project design and evaluation methodology. Skilled mediation of those differences is necessary.
- The capacity of most school districts to lead intensive professional development initiatives is hindered by a number of systemic factors, including high rates of administrator turnover, work overload of administrative and support personnel, and lack of internal expertise in the content areas or in relevant areas of research. Continued development of capacity at the local level is necessary in order to sustain school and district systemic reform.
- Although understanding of high-quality professional development has been improved by the MSP program, research-based professional development models have not emerged. A limited understanding of empirically validated theories of teacher learning and change has slowed progress toward the development of professional development models by Ohio MSP projects. While project professional development includes most features characteristic of high-quality professional development, projects usually do not conduct research on their professional development activities.
- Whole-school reform promoted by the Ohio MSP program addresses issues of reaching nonvolunteers (attendance required), but project personnel find that nonvolunteers do not participate as actively or as willingly as do volunteers. Professional development efforts require more time and focused attention to meet the needs of nonvolunteer teachers.
- All infrastructures are not equal. Early MSP efforts that relied on the existing state professional development infrastructure resulted in uneven implementation and compromised results because the infrastructure was not designed to support the initiative.

- High-quality evaluation is critical to inform continuous improvement efforts and decision making at local and state levels. Providing technical assistance to local evaluators to ensure consistency and quality is a promising approach.
- The opportunity to build upon the foundation laid by *Discovery* was invaluable to the MSP program. A large number of university faculty and teachers involved in MSP projects were formerly connected to *Discovery* in various ways.
- Systemic reform requires proactive planning for sustainability. ODE's renewed focus on sustainability has pressed MSP projects to commit time and attention to issues that impact sustainability, including partnership integration, leadership, collaboration, and intended outcomes.

Reflections for the Future

Reinventing Systemic Reform in Ohio

Systemic reform has been widely viewed as a meritorious approach to improving K–12 mathematics and science education (Clune, Porter, and Raizen 1999). Beginning with Ohio's SSI *Discovery*, funded by the NSF in 1991, and continuing today with the U.S. Department of Education's MSP Program, Ohio has used both top-down and bottom-up approaches to reform that have impacted every level of the mathematics and science education system.

The model of systemic reform described by Clune (1998) suggests that the most effective reforms are characterized by both breadth and depth of approach. Early decisions by the *Discovery* leadership to limit the project to middle school teachers, to target only 41 of 614 school districts, to initially focus on a single content area, and to require a six-week commitment from teachers enhanced the depth of the reform. As the initiative matured, *Discovery*'s reform efforts led to greater alignment among state education policies by providing information and outreach to state policy makers and education stakeholders, thereby extending the reform across the state's education system. Similarly, the Ohio MSP Program achieved breadth and depth of reform. Initially, Ohio's MSP projects broadly addressed pressing statewide needs of teachers and schools that were related to the implementation of new ACS. The first MSP statewide mathematics and science professional development projects reached each region of the state and served a very large number of teachers. Subsequent iterations of the program narrowed the focus and increased the depth of reform by targeting entire schools or small school districts, limiting content focus, and requiring teachers to commit to multiple years of professional development. Both waves of reform developed the capacity of Ohio education leaders, college and university faculty, and teachers of mathematics and science to impact the system at all levels through collaborative action.

Three decades of systemic reform efforts in Ohio have produced a number of positive outcomes for mathematics and science teachers, their students, and the state. Findings of independent evaluations of *Discovery* and of the MSP program suggest that teacher instructional practices were improved by their participation in the professional development experiences. In addition, findings from the Bridging the Gap study (Kahle, Rogg, and Tobin 1996) indicate that student achievement was enhanced after teacher participation.[9] This finding is supported by research suggesting that the mathematics and science achievement of students is influenced by the content and pedagogical knowledge of their teachers (Monk 1994; Schroeder et al. 2007; Wise 1996). Further, researchers (Desimone et al. 2002; Guyton, Fox, and Sisk 1991; Lawrenz 1975) generally concur that teacher attitudes regarding teaching and learning science impact a teacher's ability to teach standards-based science effectively. Teachers participating in three decades of Ohio's projects have reported significant positive changes in their expertise in teaching standards-based and inquiry science (Kahle, Meece, and Scantlebury 2000; Woodruff et al. 2010).

In Ohio, mathematics and science education reform has been a continuous process that reflects the dynamic nature of the educational and political systems in which teaching and learning are embedded. Progress has been alternately limited or sweeping, slow or fast, supported or challenged. Over more than 30 years, the collaborative and coordinated efforts of Ohio's mathematics and science personnel at both of the state agencies, higher education faculty, and, especially, K–12 teachers have produced an aligned system of high-quality assessments; rigorous ACS; intensive, standards-based, inservice teacher professional development; demanding standards for teacher licensure; and high expectations for student academic achievement. These sustained and diligent efforts have been rewarded with a system that supports Ohio's teachers and students in immeasurable ways.

Acknowledgments

We would like to thank Steve Meiring and Terry McCollum for their input into this chapter and, especially, for their leadership in Ohio's reform of science and mathematics education.

References

Arons, A. B. 1990. *A guide to introductory physics teaching*. New York: John Wiley and Sons.

Clay, M. M. 1993. *Reading recovery: A guidebook for teachers in training*. Portsmouth, NH: Heinemann.

Clune, W. H. 1998. *Toward a theory of systemic reform: The case of nine NSF Statewide Systemic Initiatives*. Research monograph no. 16. Madison, WI: University of Wisconsin–Madison, National Institute for Science Education.

Clune, W. H., A. C. Porter, and S. A. Raizen. 1999. Systemic reform: What is it? How do we know? *Education Week* 19 (5): 31.

Desimone, L. M., A. C. Porter, M. S. Garet, K. S. Yoon, and B. F. Birman. 2002. Does professional development change teachers' instruction? Results from a three-year longitudinal study. *Educational Evaluation and Policy Analysis* 24 (2): 81–112.

Feder, M., T. J. Horwood, and S. B. Woodruff. 2006. Using evaluation findings to build capacity for the implementation of a statewide teacher professional development program. Paper presented at the annual meeting of the American Evaluation Association, Portland, OR.

9. Also in 2008, a separate MSP solicitation was issued exclusively to fund coaching projects in mathematics. One project was funded; it was administered and evaluated independently from others funded in that year.

Freundlich, N. 1998. From Sputnik to TIMSS: Reforms in science education make headway despite setbacks. *Harvard Education Letter* 14 (5): 1–4.

Guskey, T. R., and K. S. Yoon. 2009. What works in professional development? *Phi Delta Kappan* 90 (7): 495–500.

Guyton, E., M. C. Fox, and K. A. Sisk. 1991. Comparison of teacher attitudes, teacher efficacy, and teacher performance of first-year teachers prepared by alternative and traditional teacher education programs. *Action in Teacher Education* 13 (2): 1–9.

Kahle, J. B. 2008. Systemic reform: Research, vision, and politics. In *The handbook of research on science education*, ed. S. K. Abell and N. G. Lederman, 911–934. Mahwah, NJ: Erlbaum.

Kahle, J. B., and A. Damnjanovic. 1998. *A pocket panorama of Ohio's systemic reform, 1998*. Oxford, OH: Miami University.

Kahle, J. B., J. L. Meece, and A. Damnjanovic. 1999. *A pocket panorama of Ohio's systemic reform, 1999*. Oxford, OH: Miami University.

Kahle, J. B., J. Meece, and K. Scantlebury. 2000. Urban African-American middle school science students: Does standards-based teaching make a difference? *Journal of Research in Science Teaching* 37 (9): 1019–1041.

Kahle, J. B., and S. R. Rogg. 1995. *A pocket panorama of the Landscape Study, 1995*. Oxford, OH: Miami University.

Kahle, J. B., and S. R. Rogg. 1996. *A pocket panorama of the Landscape Study, 1996*. Oxford, OH: Miami University. (ERIC Document Reproduction no. ED 419 687)

Kahle, J. B., and S. R. Rogg. 1997. *A pocket panorama of the Landscape Study, 1997*. Oxford, OH: Miami University.

Kahle, J. B., S. R. Rogg, and K. G. Tobin. 1996. *Bridging the Gap: Equity in systemic reform*. Proposal accepted by National Science Foundation, Washington, DC.

Lawrenz, F. 1975. The relationship between teacher characteristics and student achievement and attitude. *Journal of Research in Science Teaching* 12 (4): 433–437.

Lindberg, D. H. 1990. What goes 'round comes 'round doing science. *Childhood Education* 67 (2): 79–81.

Loucks-Horsley, S., J. G. Brooks, M. O. Carlson, P. J. Kuerbis, D. D. Marsh, and M. J. Padilla. 1990. *Developing and supporting teachers for science education in the middle years*. Andover, MA: National Center for Improving Science Education.

McDermott, L. C., P. S. Shafer, and M. L. Rosenquist. 1996. *Physics by inquiry: An introduction to physics and the physical sciences*, vol. 2. New York: John Wiley and Sons.

Monk, D. H. 1994. Subject area preparation of secondary mathematics and science teachers and student achievement. *Economics of Education Review* 13 (2): 125–145.

Narode, R., M. Heiman, J. Lochhead, and J. Slomianko. 1987. *Teaching thinking skills: Science*. Washington, DC: National Education Association.

National Council of Teachers of Mathematics (NCTM). 1989. *Curriculum and evaluation standards for school mathematics*. Reston, VA: NCTM.

Oches, B., H. Raffle, S. B. Woodruff, and D. Zorn. 2010. *Ohio Mathematics and Science Partnership Program cross-project evaluation: Phase 1 final report, October 2010*. Oxford, OH: Miami University, Ohio's Evaluation & Assessment Center for Mathematics and Science Education.

Ohio Department of Education (ODE). 2002. *Academic content standards K–12 mathematics*. Columbus, OH: Ohio Department of Education.

Ohio Department of Education (ODE). 2003. *Academic content standards K–12 science*. Columbus, OH: Ohio Department of Education.

Rosebery, A. S., B. Warren, and F. R. Conant. 1989. *Making sense of science in language minority classrooms.* BBN tech. rep. no. 7306. Cambridge, MA: Bolt, Beranek, and Newman. (ERIC Document Reproduction no. ED 326 059)

Schroeder, C. M., T. P. Scott, H. Tolson, T. Y. Huang, and Y. H. Lee. 2007. A meta-analysis of national research: Effects of teaching strategies on student achievement in science in the United States. *Journal of Research in Science Teaching* 44 (10): 1436–1460.

Harvard Graduate School of Education. 1998. Ohio reform shows results. *Harvard Education Letter* 14 (5): 3.

U. S. Department of Education. 2008. *ESEA: Mathematics and science partnerships (OESE), FY 2008 program performance report*. Washington, DC: U.S. Department of Education.

Wise, K. C. 1996. Strategies for teaching science: What works? *Clearing House* 69 (6): 337–338.

Woodruff, S. B., H. L. Hung, and L. Seabrook. 2009. Exploratory cluster analysis: Variability and commonality of the implementation and impact of Ohio Mathematics and Science Partnership (OMSP) Projects. Panel presentation at the annual meeting of the American Evaluation Association, Orlando, FL.

Woodruff, S. B., Y. Li, and H. C. Kao. 2010. *Evaluation of Ohio mathematics and science partnership program: Instrument validity and reliability study and study of programmatic and project-level effect size*. Oxford, OH: Miami University, Ohio's Evaluation and Assessment Center for Mathematics and Science Education.

Woodruff, S. B., T. L. McCollum, Y. Li, and N. U. Bautista. 2010. Enhancing elementary teachers' content and pedagogical knowledge through sustained professional development. Paper presented at the Annual International Conference of the National Association for Research in Science Teaching, Philadelphia, PA.

Zorn, D., L. Seabrook, J. Marks, J. Chappell-Young, S. Hung, M. Marx, and S. Woodruff. 2009. *The Ohio Mathematics and Science Partnership program external evaluation: Year 2 final report*. Oxford, OH: Miami University and University of Cincinnati, Ohio's Evaluation and Assessment Center for Mathematics and Science Education.

Zucker, A. A., P. M. Shields, N. E. Adelman, T. B. Corcoran, and M. E. Goertz. 1998. *Statewide Systemic Initiatives program*. Menlo Park, CA: SRI International.

Chapter 7

Improving and Sustaining Inquiry-Based Teaching and Learning in South Carolina Middle School Science Programs

Jeff C. Marshall, Michael J. Padilla, and Robert M. Horton

This chapter details a professional development effort that is designed to increase teacher performance and student achievement in middle school science classrooms in South Carolina. Specifically, the intervention is used to improve the quality of inquiry-based instruction in classrooms and then sustain this transformation over time. Simply put, the research results show that as the quality of inquiry-based instruction increases, student achievement likewise increases.

Nationally, we know that student performance in middle school science is in a perilous state. Only one-third of students in the eighth grade achieve a proficient or higher rating in science on key national assessments such as the National Assessment of Educational Progress (Grigg, Lauko, and Brockway 2006; NCES 2009, 2011). A major reason for this underachievement is that teachers who lead these students often lack the content knowledge and pedagogical skills needed to address the crisis effectively. When teacher instructional practice improves, student achievement also increases (Heller et al. 2012; Marzano, Pickering, and Pollock 2001; Stronge 2002).

South Carolina's science curriculum standards are rated as excellent and among the most demanding in the nation (Gross et al. 2005)—yet student achievement continues to languish, and state assessment results mirror the poor national data. Further, our research shows that both the quantity and quality of inquiry instruction are inadequate (Marshall et al. 2009). The professional development intervention featured in this chapter details the efforts to transform teacher practice to higher-quality inquiry-based instruction that is focused on key content standards.

Rationale

The goal of the intervention, which is part of a systemic initiative entitled Inquiry in Motion (IIM), is straightforward—transform middle school science teaching by improving the quality of inquiry-based instruction, thus improving student achievement. This work has led to dramatic increases in both teacher quality and student achievement, while simultaneously reducing the achievement gap that exists along racial and socioeconomic lines (Marshall, Horton, and Smart 2009; Marshall, Horton, and White 2009; Marshall and Horton 2009).

Even though many professional development and curricular efforts strive to improve the quality of inquiry-based instruction, it is clear that teachers, researchers, and educational leaders often assume that all inquiry is of equivalent quality. The findings from the intervention suggest that this is not so; specific tools that were designed to work in conjunction with the intervention, when combined with individualized support, help teachers achieve a proficient level of inquiry-based instruction. Further, a plan for continuing and sustaining instructional improvement is often missing when changes are made. The intervention, guided by specific tools described in this chapter, lasts a minimum of two years, and when the research team leaves a school, they leave behind a cohort group of teachers determined to maintain the transformation.

One of the great challenges of professional development interventions is to assert accurately and validly that subsequent change in student achievement can be attributed to changes made in teacher performance. By using a validated national test with multiple controls to minimize confounding variables, we are able to provide a clear understanding of teacher proficiency based on student achievement relative to the IIM intervention.

The IIM professional development intervention provides a comprehensive model for improving the quality of inquiry-based instruction that is supported by the research literature and by past research efforts. Specifically, we have seen improvement in student achievement when teacher performance is enhanced via a variety of support mechanisms that ultimately attempt to improve student achievement.

Model

Structure of the IIM

To transform teacher practice by improving participants' inquiry-based practice, two professional development institutes (PDIs) and a set of support structures have been established. These are discussed in detail in this section.

PDIs

In order for transformation of practice to occur and endure, teachers need sustained professional development experiences with sufficient support (Loucks-Horsley et al. 2003; Supovitz and Turner 2000). The two-year-long PDIs provide the framework of this sustained model in an effort to achieve transformative practice. Collectively, the PDIs are designed to improve the quality of inquiry instruction among teachers via rich, transformative experiences and ongoing support. Ongoing support helps to ensure that the changes are sustained over time (Bransford, Brown, and Cocking 2000; NRC 2000).

During the last four years, the following lessons were learned in an effort to improve teacher practice and student achievement: (1) Transformation of teaching practice is more likely to occur when individual schools, not entire districts, are targeted. Not only does this promote the development of a community within the schools, but also the capacity and energy of the research team are focused, not dispersed. Specifically, two schools per year should be targeted. (2) Follow-up meetings need to be frequent and must provide personalized classroom support. (3) An advanced or second-year PDI is necessary to support and promote sustained transformation after initial growth has occurred. The expectation is that at least 40–50% of the original participants will continue as teacher leaders during the second year.

Annually, two levels of PDIs are offered—PDI-1 for first-year participants and PDI-2 for second-year participants. Both levels of PDIs are aligned with well-regarded research concerning transforming teaching practice (Banilower et al. 2006; Loucks-Horsley et al. 2003; Supovitz and Turner 2000). The logic model for our work is provided in Table 7.1 (p. 106). First, district leaders help identify two schools they believe are most critical to target. Once a school has been identified, the research team needs to meet with building administrators to describe the institutes. If the administrators are interested and supportive, then a meeting with the teachers can be scheduled; a minimum of 60% of the science teachers must agree to participate in PDI-1 before the school becomes a partner. This process ensures both administrative and teacher support.

PDI-1

Involvement in PDI-1 includes a two-week summer commitment, four group follow-up sessions, and individual interactions with the research team during the academic year. During the 60 hours of summer commitment, the approximately 20 annual participants (1) learn to use and then apply the 4E × 2 Instructional Model (described later) as they create exemplar lessons, which are content-rich standards-based inquiry lessons; (2) learn to use EQUIP (Electronic Quality of Inquiry Protocol; described later) to guide the development of an explicit action plan that charts steps necessary to transform their instructional practice; and (3) work to improve any content deficiencies for units of focus.

Four follow-up sessions, collectively totaling 12 hours, are held during the academic year to provide further support as transformation of practice occurs. During follow-up sessions, participants (1) discuss successes and challenges central to planning and implementing inquiry-based lessons, (2) discuss growth achieved relative to indicators identified from the EQUIP instrument, (3) work collectively to improve major instructional and management issues that surface, and (4) modify previously developed exemplars and develop an additional exemplar as a team. Additionally, during the academic year, each teacher's classroom is visited at least four times for two purposes. One purpose is geared toward research: EQUIP is used to observe teacher performance relative to assessing the quality of inquiry-based instruction. This allows for the tracking of growth in teacher performance and to correlate teacher performance with student performance. The second purpose of these visits is to provide individualized support for participants. This support takes any of the following forms: teach while the participant observes; coteach with the participant; observe and then engage the participant in postinstruction conversations and reflections; teach while the participant observes another participant; or provide technical support (e.g., show how to use electronic sensors to study motion).

Table 7.1 Logic models for PDI-1 and PDI-2

	Inputs	Activities	Outputs	Outcomes	Impacts
PDI-1	Improve inquiry-based instruction in high-need schools Improve teacher effectiveness in using inquiry-based instruction Improve student achievement	Engage university research teams, district administrators, schools, teachers, teacher collaborators, and resources Have research team meet with district personnel to identify high-need schools Meet with principals and teachers Summer PDI-1, follow-up professional development sessions, class visits for personalized support Surveys and focus groups	Schools and teachers are identified At least 60% of science teachers at a school are involved in intervention Teachers learn to apply the 4E × 2 Instructional Model, dynamic WebTool, and EQUIP Teachers develop and revise exemplars Assess and track individual teacher and school progress Examine student work and academic growth	Increased district and school support for inquiry-based teaching Improved teacher instruction using inquiry Increased use of technology Increased collaboration among teachers Teachers produce exemplars Ongoing research Demonstrated proficiency using EQUIP to transform practice Measured student growth relative to a virtual comparison group	Greater use of and greater quality of inquiry-based instruction Improved student achievement
PDI-2	Create teacher leader and build support structure for inquiry instruction	Converse with school leadership to select teachers from PDI-1 to become teacher leaders	Select and train PDI-2 teachers to become teacher leaders to assist with PDI-1 teachers and other teachers	Develop teacher leaders who are prepared to provide classroom support at local schools, work with building administrators, and support development of a school implementation plan	Greater use of and greater quality of inquiry-based instruction Improved student achievement Impact is both at the teacher and at the school level

PDI-2

The advanced leadership institute for second-year participants, PDI-2, prepares teachers to become school leaders as their practice continues to be transformed and internalized. For each PDI-2, 8 to 10 participants are recruited from the previous year's PDI-1 cohort. Selection is based on demonstrated excellence in inquiry instruction (determined by EQUIP ratings), evidence of leadership potential (how the participants have helped to support others during their first year), and the recommendation and support of a building administrator. PDI-2 participant commitment involves two weeks during the summer, four group follow-up meetings with PDI-1 participants, monthly meetings at their home school during the ensuing academic year, and individual interactions each academic quarter with members of the research team.

Offered annually, PDI-2 strives to support the transformation of new PDI-1 participants from the two new partnering schools and to continue to improve and sustain the transformation of teachers in their home schools. PDI-2 teachers provide support to PDI-1 teachers as they create standards-based inquiry lessons. They also help PDI-1 teachers address and overcome any obstacles they may encounter while implementing inquiry-based instruction. To sustain transformation in their own schools, PDI-2 participants develop and implement a school plan. During the summer component of PDI-2, a building administrator must attend at least a half-day to assure buy-in. As one critical component of this plan, PDI-2 participants lead monthly meetings in which the previous year's PDI-1 participants discuss their successes and struggles, create and/or improve inquiry-based lessons, and identify their needs to implement inquiry instruction effectively. PDI-2 participants also model effective teaching at these meetings. Such sustained intervention encourages continued progress toward transformed practice. Finally, PDI-2 participants obtain and post student work samples and video clips for their inquiry-based exemplars in order to provide more evidence for other teachers who may wish to implement them.

Formative research findings have guided revisions to PDI-1, and future research will continue to highlight areas needing revision for both PDI levels. A collective case study showed that participating teachers, while differing in skill and experience, all took significant steps in transforming their instructional practice (Marshall and Smart 2013). This research also demonstrated the need for support structures to assist all teachers in achieving a more sustainable transformation to inquiry-based instruction.

Support Structures

Several key support structures, designed for both PDI-1 and PDI-2 participants comprise an integral part of the professional development intervention—all are supported by theory and practice. The supports that are innovative and were created specifically for the IIM program include (1) the 4E × 2 Instructional Model, (2) the dynamic WebTool (including exemplars), and (3) EQUIP. These three supports are continually reinforced during summer interactions, follow-up meetings, face-to-face interactions, and by members of the cohort. These support structures are discussed below.

Support Structure #1: 4E × 2 Instructional Model

Student achievement increases when teachers effectively incorporate three critical learning constructs into their teaching practice: (1) inquiry instruction (NRC 2000, 2012), (2) formative assessment (Black and Wiliam 1998a; Marzano 2006), and (3) teacher reflection (NBPTS 2006). The 4E × 2 Instructional Model, which provides the instructional framework for IIM, integrates these learning constructs into a single dynamic model that guides the transformation of instructional practice (Marshall, Horton and Smart 2009). We address these three constructs in the paragraphs that follow.

Inquiry instruction provides a solid and essential strategy for learning science (Bransford, Brown, and Cocking 2000). Currently, in the BSCS 5E Instructional Model (Engage, Explore, Explain, Elaborate, and Evaluate) that has been widely adopted by many science educators (Bybee, Taylor, and Gardner 2002), the fifth E (Evaluate) is often perceived as summative, not formative, and thus left to the end of instruction. Also, teacher reflection is not explicitly integrated and thus is often neglected. The 4E × 2 Instructional Model adopts the core of the 5E Model while integrating and explicitly linking formative assessment and teacher reflection to *each* phase of inquiry instruction as shown in Figure 7.1.

Figure 7.1

FRAMEWORK FOR THE 4E × 2 INSTRUCTIONAL MODEL

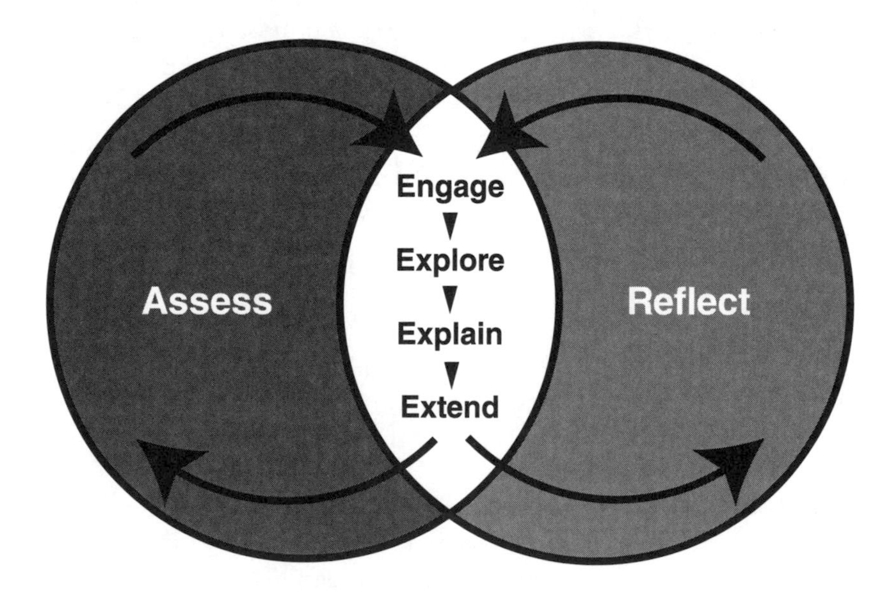

Formative assessment of learning, when used effectively, helps teachers make instructional decisions that lead to deeper understanding for all students. Thus, successfully incorporating formative assessment helps narrow the achievement gap (Black and Wiliam

1998a, 1998b). When formative assessment is integrated into each component of the inquiry learning process, teachers become more intentional and more effective in their practice. For example, teachers can provide a formative probe, a short scenario that identifies prior knowledge and misconceptions during the Engage portion of the lesson, or they can pause after every three PowerPoint slides to have students summarize the key ideas during the Explain part of the lesson.

Teacher reflection is a process of explicitly examining instructional practice and making informed decisions about the trajectory of instruction because it makes teachers more aware of students' understandings. Such reflection encourages practice that is more responsive to students' needs, thus stimulating deeper learning. The National Board for Professional Teaching Standards uses reflective practice as the central tenet to its certification process (NBPTS 1994, 2000, 2006). The 4E × 2 Instructional Model directly integrates reflective practice into each step of the inquiry instructional process.

Support Structure #2: WebTool (including exemplars)

IIM's WebTool is designed to encourage, guide, and then maintain the desired teacher transformations as we strive for a higher quality of inquiry-based curriculum and instruction. Specifically, the WebTool (free to all registered users) allows teachers to view exemplars that other educators have created, modify existing exemplars to meet individual needs, create new exemplars using the online template, and share exemplars with other teachers. Even though the WebTool uses the 4E × 2 Instructional Model, it is versatile enough to accommodate other models such as the Learning Cycle or the 5E Instructional Model. The WebTool can be found by selecting the lesson planning tab at the *www.clemson.edu/iim* website.

Each exemplar contains detailed guidelines for each of the four E's of inquiry instruction, sound strategies for formative assessment, and guiding questions for teacher reflection, thus ensuring that the lesson is complete in all areas. Further, the website helps guide teachers through the process of creating coherent, well-aligned lessons or units by providing boxes (many with drop-down menus) for things such as lesson overview, standards, materials, teacher support documents, description of each component of inquiry, and assessments that will be used during each component of inquiry. PDI-1 teams, typically consisting of four teachers, create, implement, and refine two exemplars by the end of the academic year (one is created during the summer and the other during the academic year). In an attempt to develop rigorous inquiry-based exemplars, every team has the support from PDI-2 participants, a graduate student, and educational specialists.

The creation, implementation, refinement, evaluation, and dissemination of exemplars is part of an iterative process that aligns with the Japanese Lesson Study, which emphasizes teacher collaboration in creating, testing, and refining lessons based on student learning (Lewis 2002; Puchner and Taylor 2006; Stigler and Hiebert 1999). Thus, exemplars focus on both the processes and products associated with transforming teacher practice. Exemplars are modified after implementation based on the teachers' reflections and project leaders' analysis and recommendations.

To assist teachers in making their instruction more intentional, student work samples are being added to the IIM website to illustrate unsuccessful, moderately successful, and highly successful student performance. Further, teachers are beginning to post videos that feature key portions of exemplars during instruction. When completed (though they are considered complete without student work samples and video illustrations), all exemplars are posted and made available to the public via the WebTool. Although the WebTool allows any teacher to post an exemplar for public viewing, only those that have been reviewed by content and educational specialists and meet the established criteria are given the silver-level (18 of 20 posted components) or gold-level (20 of 20 components) endorsement. Collectively, these endorsements, student work samples, and video illustrations of instruction provide information about the quality of the exemplar from both curricular and instructional perspectives.

The dynamic WebTool has evolved from four years of development and has been piloted with over 100 elementary, middle, and high school science and math teachers. The interface component allows site visitors to view existing exemplars. It also allows registered users to (1) modify an exemplar by copying it to My WorkSpace or (2) create a new lesson guided by an online template based on the 4E × 2 Instructional Model. A new lesson (exemplar) may either be kept private or posted for public use. Further, the interface allows teachers to interact and work together in asynchronous ways. Finally, the interface's Administrative area allows project leaders to monitor quality, post endorsements, communicate with exemplar authors, and track database usage.

The current WebTool provides a rich, interactive support structure for teachers who may or may not be participants in our professional development efforts. As of this writing, the WebTool has largely been used by teachers in South Carolina, but we have had visitors to the site from 73 countries and all 50 states. Further, through the use of Google analytics, we are able to see that visitors to the WebTool spend on average more than 10 minutes on the site, which indicates a high degree of activity and engagement even with little promotional encouragement to this point. The primary advantage of this site over other lesson planning sites is the dynamic nature, which allows authors to edit lessons, create new lessons, and add the lessons of others to their own My Workspace area to be modified as necessary. By spring 2014, all lessons will be aligned to the *Next Generation Science Standards* (NGSS Lead States 2013) and be easily edited to address different standards that are used in some states.

Support Structure #3: EQUIP

EQUIP provides a powerful guide for assessing the quantity and quality of inquiry instruction being implemented by PDI participants (Marshall et al. 2008). EQUIP provides teachers, researchers, and educational leaders with both a macro and micro look at inquiry instructional practice.

EQUIP was derived primarily from two heavily researched observation protocols (Henry, Murray, and Phillips 2007): "The Reformed Teaching Observation Protocol" (Sawada et al. 2000) and "Inside the Classroom: Teacher Interview Protocol" (Horizon

Research 2002). Though highly reliable and often used by others, neither protocol met our needs for an instrument that is focused explicitly on inquiry-based instruction and contains distinct, descriptive rubrics (Marshall, Horton, and White 2009; Marshall, Smart, and Horton 2010). Feedback from education experts, content experts, and our pilot group of teachers has established the content validity of EQUIP (N = 102). Our research shows that EQUIP also has high internal consistency and strong inter-rater reliability. More importantly, four constructs (discussed below) with high validity have emerged through a confirmatory factor analysis.

EQUIP uses descriptive rubrics to assess 19 total indicators for four key constructs:

- Instruction: 5 indicators on instructional practice (e.g., teacher role, order of instruction)
- Discourse: 5 indicators on discourse (e.g., questioning ecology, classroom interactions)
- Assessment: 5 indicators on assessment (e.g., prior knowledge, assessment type)
- Curriculum: 4 indicators on curriculum (e.g., content depth, standards)

During the class, a snapshot view is taken every five minutes. Then, the indicators for each construct are assessed after the observation and combined to determine an overall score for the construct. EQUIP can be found at *iim-web.clemson.edu/?page_id=166.*

EQUIP serves both as a research instrument and as a tool that provides a solid formative assessment to help teachers target specific areas of growth needed in their instructional practice. The value and wide acceptance of this instrument has been evident during the last three years. It has specifically resulted in publications in both research and practitioner journals, presentations at practitioner and researcher conferences, and consulting projects with schools and universities.

Both PDI interventions, coupled with the aforementioned support structures, provide teachers and schools with a sustainable support mechanism to achieve high-quality inquiry instruction and improved student achievement. Further, our research indicates that when the 4E × 2 Model, the WebTool, and EQUIP are used as the foundational strategy for professional development, instruction is transformed and student learning increases. Specifically, data collected show the following: critical thinking among students increases (Marshall and Horton 2011), and both science content and process knowledge increases as measured by the Measures of Academic Progress (MAP) tests. Further, the students in the classes of participating teachers outperform (1) students from prior years that had the same teacher, (2) students throughout the district (control for district effects), and (3) similarly classified students from a virtual control group (control for student demographics). The increase in student performance was determined by analysis of data provided by a study with the Northwest Evaluation Association (NWEA) and has not been published yet. The framework to support the method used in this study has been published (Cronin, Marshall, and Xiang 2009).

Lessons Learned

Research surrounding the IIM interventions is focused on three issues: (1) Are teachers transforming their practice relative to inquiry-based instruction? (2) Are changes sustained and does growth continue? (3) Does student achievement increase as a result of teacher transformation? To address these questions, we collected and analyzed data from classroom observations and interactions (EQUIP and conversations), surveys (pre- and postintervention), focus groups (during follow-up sessions), posted lessons (WebTool), and student achievement (MAP testing pre- and postintervention). We discuss all of these briefly below, and then focus in greater detail on our findings in regard to student achievement.

Classroom Observations

The quantity and quality of inquiry instruction implemented by PDI participants is measured using EQUIP (Marshall et al. 2008). In the spring prior to PDI-1 involvement, every participant is observed to establish a baseline performance. During the academic year, each PDI-1 and PDI-2 participant is observed at least once every academic quarter. Collectively, the observations made using EQUIP allow researchers to assess teachers' transformations over time.

Four levels of inquiry instruction were established: pre-inquiry (level 1), developing (level 2), proficient (level 3), and exemplary (level 4). EQUIP was written so that level 3 is aligned with the targeted goals laid forth by the science and math standards. The desired outcome for PDI-1 and PDI-2 participants is for them to have demonstrated proficiency (level 3) or above in the four constructs (instruction, discourse, assessment, and curriculum) and on the Overall Lesson Summary assessment measured by EQUIP. Overall, typical growth averages just less than one full level on the Overall Lesson Summary over the course of the first year of intervention, with continued growth during the second year of participation. PDI-2 participants' growth more frequently is seen being transferred to all teaching experiences, not just the teaching of inquiry-based lessons that were prepared with their team. EQUIP data also show the individual indicators on which participants struggle so that they can be targeted during follow-up meetings and in subsequent training for both PDI-1 and PDI-2 participants.

Survey

To better understand the degree of transformation that has occurred, an existing survey is used to measure (1) the quantity and quality of inquiry-based instruction that teachers self-report, (2) teachers' motivation and self-efficacy related to inquiry instruction, and (3) their understanding of what inquiry is and what it looks like in their classroom. PDI-1 and PDI-2 participants are surveyed at two points during the year, at the beginning of PDI-1 (diagnostic) and at the end of the school year.

The survey is based on the highly reliable instrument that was completed by 1,222 Greenville County (SC) mathematics and science teachers in spring 2007 (Marshall et al. 2009). Likert-scaled items explore motivation, self-efficacy, and participants' views of the actual and ideal amounts of time spent on inquiry instruction. Free-response items ask teachers to demonstrate their understanding of inquiry instruction and explain the role of an effective teacher in such a setting. By studying results longitudinally, one is able to gain insights into the effectiveness of the PDIs, exemplars, and the dynamic WebTool in transforming practice. Multiple administrations provide both formative and summative feedback.

Results from previous years indicate that more than 90% of the participants believe their ability to plan and implement inquiry-based lessons has increased significantly and that student achievement has increased due to their enhanced abilities with inquiry-based strategies.

Exemplars

Of the completed exemplars, 80% achieve a high-quality rating according to the Lesson Quality Checklist Rubric (Figure 7.2, p. 115) that we use to analyze them. In the last two years, 16 out of 20 (80%) of the posted exemplars have earned at least a silver rating. The administrative features of the WebTool coupled with Google Analytics provides the ability to track where and how often exemplars are accessed by web users.

Focus Groups

During a fall and spring follow-up meeting for PDI-1 and PDI-2 participants, focus groups are held to assess the successes and challenges that exist relative to achieving a sustainable plan in their schools. The findings from these groups provide formative feedback to allow more support to be provided where necessary for each participant. Patterns within the partner schools can be determined to see whether schoolwide adjustments are necessary. These focus groups have consistently supported the findings that participants believe their knowledge of and ability to implement inquiry-based instruction has grown and that their students are learning at deeper levels due to this increased proficiency with inquiry.

Student Achievement

Student growth is analyzed using the MAP, a reliable and valid assessment by NWEA (2004) that is used by schools in 48 states. As an adaptive test, MAP provides either more or less challenging items, depending on students' success or failure on previous questions. Further, because it is aligned with state and national science standards (NWEA 2005), MAP can pinpoint students' current level of achievement. Students are assessed both in the fall and in the spring; thus, growth during the academic year can be readily determined. This test also allows success to be studied with various ability levels, thus providing a means to research possible effects on the achievement gap.

MAP has several inherent strengths. First, because test items are aligned to state science standards, it has high predictive validity when compared to other state assessments (Cronin et al. 2007; NWEA 2005). Second, MAP allows the performance of a given teacher's students to be studied at both the macro level (e.g., science content and process knowledge scores) and micro level (e.g., weather, forces, ecosystems), which are targeted by exemplars without requiring an additional test. Third, because it is adaptive, MAP provides a broader, more robust sample of the entire domain than a fixed-form test does (NWEA 2003). Further, since all partner districts already use MAP, no additional testing is necessary to obtain a reliable and valid measure of student performance. Finally, NWEA is piloting test items that are aligned to the *Next Generation Science Standards* (NGSS Lead States 2013) beginning in fall 2013.

Two MAP tests are relevant to this project: science content and science process. These tests are used to measure student growth in academic performance and serve as an outcome measure for the proposed interventions.

To assess the effectiveness of the project, we compare growth among three groups of students: (1) the study group comprising students taught by PDI-1 and PDI-2 teachers, (2) a comparison group comprising students taught by teachers in the same district who have not participated in PDI-1, and (3) a virtual comparison group (VCG) comprising students drawn from NWEA's Growth Research Database who are matched to study-group students. To form the VCG, each study group student is matched with 21–51 students who score within one point on the test taken in the fall, are tested within the same seven-day window, attend a school that has the same rate of free and reduced lunch (within 5%), and possess the same urban or rural, sex, and ethnic designations.

Our quasi-experimental design controls for several important factors that often confound educational studies:

- Pre- and postintervention measurement of the study group helps control for prior performance of study group participants.
- Pre- and postintervention measurement of the students of nonparticipating educators helps control for a school or school system effect on the growth of students.
- The use of a VCG helps control for effects that might be a product of variance within the student cohorts. Consequently, any gains reported by the program are linked to improvements in instruction.

By collecting data for multiple years, including a year after completion of the PDI-1, we can determine whether any effect generated by the program persists or whether there is J-curve effect associated with this intervention. The J-curve phenomenon suggests that, as a new reform is implemented, a lag, or even a slight drop, can be expected until the teacher becomes comfortable with the changes (Erb and Stevenson 1999). If the reform is effective, student outcomes will improve in the long run, provided that sufficient time is allowed to overcome the J-curve effect (Yore, Anderson, and Shymansky 2005).

Figure 7.2

LESSON CHECKLIST FOR DETERMINING GOLD- AND SILVER-QUALITY LESSONS

CRITERIA	MEETS CRITERIA	DOESN'T MEET CRITERIA
Essential (All criteria must be met to receive gold or silver certification)		
National standards are clearly specified and aligned with state standards.	☐	☐
Sufficient background information is provided.	☐	☐
The instructional plan effectively addresses the national and state standards specified for this lesson.	☐	☐
Sufficient detail is given for each phase of the instructional plan to allow another teacher to duplicate the lesson.	☐	☐
Explore phases of the instructional plan precede explanations of content.	☐	☐
Lessons are largely student-focused with students taking an active role in learning.	☐	☐
Real world or meaningful context is embedded in the lesson to promote conceptual understanding.	☐	☐
Safety issues are addressed as necessary.	☐	☐
The lesson addresses a fundamental concept or big idea in either math or science.	☐	☐
Lesson is cohesive: standards, lesson, and assessments well aligned.	☐	☐
Important (Gold = all 10 criteria met; Silver = at least 8 criteria met)		
A concise lesson overview is provided.	☐	☐
State standards are indentified by number and description.	☐	☐
Prerequisite knowledge is addressed as necessary.	☐	☐
Necessary materials are listed.	☐	☐
All necessary teacher support documents are present and labeled appropriately.	☐	☐
The Instructional Plan includes engage, explore, explain, and extend.	☐	☐
At least one subcomponent (e.g., prior knowledge) is specified for each phase of the instructional plan.	☐	☐
Formative assessments that will guide instruction are provided and described for each phase of the instructional plan.	☐	☐
Questions for teacher reflection are present for each phase of the instructional plan.	☐	☐
The lesson makes connections to other concepts in math or science.	☐	☐

Note: For exemplars to earn gold distinction, all criteria must be met. For exemplars to earn silver distinction, all essential criteria and at least 8 out of 10 of the important criteria must be met.

Our analysis, conducted in partnership with NWEA, has shown that at the macro level (science content and process knowledge) students of PDI-1 participants outperform both the students of district teachers and those in the VCG ($p < .05$). Though we do have evidence that performance continues to grow, we do not yet have sufficient data to generalize our results. Further, the preliminary data also suggest that the teacher performance on EQUIP is highly predictive of the growth on science content and science process. These data should still be considered tentative until more data from several different schools have been gathered, but the average EQUIP score for the discourse construct shows that the EQUIP discourse scale accounts for approximately 72% of the variance seen on the MAP science content ($p < .05$). This suggests how vital developing solid classroom discourse is to academic achievement. Additionally, the EQUIP score on the assessment construct score accounts for 59% of the variance seen on the MAP content test ($p < .05$). If this trend continues to hold true, this suggests that inquiry-based instruction may be beneficial to improving both content and process achievement in the schools. Little research is available addressing both aspects of learning. Inquiry-instruction research often focuses exclusively on the degree to which we are able to improve critical thinking, which is not only often difficult to assess without validated instruments but may or may not be at the expense of learning key content knowledge.

Finally, our project provides data on how students of different socioeconomic and cultural backgrounds perform relative to one another. This allows us to determine to what degree the proposed interventions support the narrowing of the achievement gap. Preliminary data suggest that not only is performance of students from teachers who were in PDI-1 higher than the two comparison groups (district and VCG) overall, but that previously underperforming groups make greater progress relative to their comparison groups. Thus, we are making significant headway on narrowing the achievement gap. In summary, MAP provides a robust, well-researched instrument that measures student achievement and growth data in reliable and valid ways. It provides a cost-effective, time-efficient, flexible means to measure student achievement and growth with powerful controls. Results indicate that our interventions are making a positive difference.

Reflections for the Future

So, if a program seeks to improve teacher performance relative to inquiry-based instruction that results in improved student achievement, what portions of this program are transferrable to other professional development efforts? The 4E × 2 provides a model that teachers can adopt to frame their lessons to make them more inquiry driven. The interactive WebTool is available free of cost to the public; this not only provides a structure for developing lessons (on both an individual or team basis) but also serves as a repository for existing lessons. Finally, whether conversations take place districtwide, schoolwide, or teamwide, EQUIP provides a rich framework to guide conversations about teaching practice. By targeting one or two of the indicators at a time, teachers can explicitly discuss how they are attempting to transform their practice to a level 3 (proficient) performance in a given area. Three very specific areas that could be targeted include the following:

Lesson Plans

In teams, have teachers examine current lesson plans, reworking them if necessary to ensure that students are able to explore phenomena before the explanation of the concept occurs. This idea of Explore before Explain is at the heart of our work. Further, discuss ways to improve the role that students play during the explanation (sense-making) process. Realize that inquiry-based learning needs to be scaffolded so that learners adapt to new ways of learning. Providing very short Explore–Explain opportunities for students initially helps ensure that no student has time to become overly frustrated or left behind.

Discourse

When students become more involved in the conversations in the classroom, then they are required to think more deeply about the material—thus promoting learning. However, if students have shut down from such interactions for the last few years, then they need something to encourage them to reenter the academic arena. One suggestion is to practice three to five engaging questions that will be asked during the class period. This ensures that questioning goes beyond basic perfunctory recitation of one word answers, and it forces us as teachers to make sure that a large portion of our conversations are about engaging, thought-provoking material.

Assessment

Formative assessments have been lauded for years as critical for successful teaching and learning. However, formative assessments in inquiry-based learning have not been as thoroughly articulated. Teachers need to be provided with time to discuss and reflect on examples of what formative assessment looks like at each step in the learning process. For instance, during the Engage phase, teachers should be concerned with determining the misconceptions that students possess as well as the prior knowledge that they bring with them to the lesson. By explicitly examining the lessons being led, the discourse being facilitated, and the assessments that are given, professional development leaders are provided with some key areas to explore with teachers that begin to help transform their teaching practice.

Acknowledgments

Funding from an NSF-CAREER grant (DRL-0952160) and by the South Carolina Center of Excellence Program (Center of Excellence for Inquiry in Mathematics and Science) provides the financial support necessary for implementing and researching the PDI-1 and PDI-2 associated with IIM.

References

Banilower, E. R., S. E. Boyd, J. D. Pasley, and I. R. Weiss. 2006. *Lessons from a decade of mathematics and science reform: A capstone report for the local systemic change through teacher enhancement initiative.* Chapel Hill, NC: Horizon Research.

Black, P., and D. Wiliam. 1998a. Assessment and classroom learning. *Assessment in Education* 5 (1): 7–74.

Black, P., and D. Wiliam. 1998b. Inside the black box: Raising standards through classroom assessment. *Phi Delta Kappan* 80 (2): 139–148.

Bransford, J. D., A. L. Brown, and R. R. Cocking. 2000. *How people learn: Brain, mind, experience, and school.* Expanded edition. Washington, DC: National Academies Press.

Bybee, R. W., J. A. Taylor, and A. Gardner. 2002. *The BSCS 5E Instructional Model: Origins, effectiveness, and applications*. Colorado Springs, CO: BSCS.

Cronin, J., G. G. Kingsbury, M. Dahlin, D. Adkins, and B. Bowe. 2007. Alternate methodologies for estimating state standards on a widely-used computerized adaptive test. Paper presented at the annual meeting of the National Council on Measurement in Education, Chicago, IL.

Cronin, J., J. C. Marshall, and Y. Xiang. 2009. Assessing the effectiveness of a science and mathematics teacher development program through use of virtual comparison groups. Paper presented at the annual meeting of the National Association of Researchers of Science Teaching, Garden Grove, CA.

Erb, T. O., and C. Stevenson. 1999. Middle school reforms throw a "J-curve": Don't strike out. From faith to facts: Turning points in action. *Middle School Journal* 30 (5): 45–47.

Grigg, W. S., M. A. Lauko, and D. M. Brockway. 2006. *The nation's report card: Science 2005*. National Center for Education Statistics. *nces.ed.gov/pubsearch/pubsinfo.asp?pubid=2006466*

Gross, P. R., U. Goodenough, S. Haack, L. S. Lerner, M. Schwartz, and R. Schwartz. 2005. *The state of the state science standards.* Washington, DC: Thomas B. Fordham Institute.

Heller, J. I., K. R. Daehler, N. Wong, M. Shinohara, and L. W. Miratrix. 2012. Differential effects of three professional development models on teachers' knowledge and student achievement in elementary science. *Journal of Research in Science Teaching* 49 (3): 333–362.

Henry, M., K. S. Murray, and K. A. Phillips. 2007. Meeting the challenge of STEM classroom observation in evaluating teacher development projects: A comparison of two widely used instruments. Paper presented at the annual meeting of the American Evaluation Association, Baltimore, MD.

Horizon Research. 2002. Inside the classroom interview protocol. *www.horizon-research.com/instruments/clas/cop.pdf*

Lewis, C. 2002. Does lesson study have a future in the United States? *Nagoya Journal of Education and Human Development* 1: 1–23.

Loucks-Horsley, S., N. Love, K. E. Stiles, S. Mundry, and P. W. Hewson. 2003. *Designing professional development for teachers of science and mathematics*. 2nd ed. Thousand Oaks, CA: Corwin Press.

Marshall, J. C., and R. Horton. 2009. Developing, assessing, and sustaining inquiry-based instruction: A guide for math and science teachers and leaders. Saarbrücken, Germany: VDM Verlag

Marshall, J. C., and R. M. Horton. 2011. The relationship of teacher-facilitated inquiry-based instruction to student higher-order thinking. *School Science and Mathematics* 111 (3): 93–101.

Marshall, J. C., R. Horton, B. L. Igo, and D. M. Switzer. 2009. K–12 science and mathematics teachers' beliefs about and use of inquiry in the classroom. *International Journal of Science and Mathematics Education* 7 (3): 575–596.

Marshall, J. C., B. Horton, and J. Smart. 2009. 4E × 2 Instructional Model: Uniting three learning constructs to improve praxis in science and mathematics classrooms. *Journal of Science Teacher Education* 20 (6): 501–516.

Marshall, J. C., B. Horton, J. Smart, and D. Llewellyn. 2008. EQUIP: Electronic quality of inquiry protocol. Clemson University Inquiry in Motion Institute. *www.clemson.edu/iim*

Marshall, J. C., R. Horton, and C. White. 2009. EQUIPping teachers: A protocol to guide and improve inquiry-based instruction. *Science Teacher* 76 (4): 46–53.

Marshall, J. C., and J. B. Smart. 2013. Teachers' transformation to inquiry-based instructional practice. *Creative Education* 4 (2): 132–142.

Marshall, J. C., J. Smart, and R. M. Horton. 2010. The design and validation of EQUIP: An instrument to assess inquiry-based instruction. *International Journal of Science and Mathematics Education* 8 (2): 299–321.

Marzano, R. J. 2006. *Classroom assessment and grading that work*. Alexandria, VA: ASCD.

Marzano, R. J., D. J. Pickering, and J. E. Pollock. 2001. *Classroom instruction that works: Research-based strategies for increasing student achievement*. Alexandria, VA: ASCD.

National Board for Professional Teaching Standards (NBPTS). 1994. *What teachers should know and be able to do*. Washington, DC: NBPTS.

National Board for Professional Teaching Standards (NBPTS). 2000. *A distinction that matters: Why national teacher certification makes a difference*. Greensboro, NC: Center for Educational Research and Evaluation.

National Board for Professional Teaching Standards (NBPTS). 2006. *Making a difference in quality teaching and student achievement*. Arlington, VA: NBPTS.

National Center for Education Statistics (NCES). 2009. *The nation's report card: Science 2009. National Assessment of Educational Progress at Grade 8*. Washington, DC:NCES. *nces.ed.gov/pubsearch/pubsinfo.asp?pubid=2011451*

National Center for Education Statistics (NCES). 2011. *The nation's report card: 2009 2011 science assessments*. Washington, DC: NCES. *nationsreportcard.gov/science_2011/*

National Research Council (NRC). 2000. *Inquiry and the national science education standards: A guide for teaching and learning*. Washington, DC: National Academies Press.

National Research Council (NRC). 2012. *A framework for K–12 science education: Practices, crosscutting concepts, and core ideas*. Washington, DC: National Academies Press.

NGSS Lead States. 2013. *Next Generation Science Standards: For states, by states*. Washington, DC: National Academies Press. *www.nextgenscience.org/next-generation-science-standards*.

Northwest Evaluation Association (NWEA). 2003. *Technical manual*. Lake Oswego, OR: Northwest Evaluation Association.

Northwest Evaluation Association (NWEA). 2004. *Reliability and validity estimates: NWEA achievement level tests and measures of academic progress*. Lake Oswego, OR: Northwest Evaluation Association.

Northwest Evaluation Association (NWEA). 2005. *NWEA reliability and validity estimates: Achievement level tests and measures of academic progress*. Lake Oswego, OR: Northwest Evaluation Association.

Puchner, L. D., and A. R. Taylor. 2006. Lesson study, collaboration and teacher efficacy: Stories from two school-based math lesson study groups. *An International Journal of Research and Studies* 22 (7): 922.

Sawada, D., M. Piburn, K. Falconer, J. Turley, R. Benford, and I. Bloom. 2000. *Reformed Teaching Observation Protocol (RTOP)*. Tech. rep. no. IN00-01. Tempe, AZ: Arizona State University.

Stigler, J. W., and J. Hiebert. 1999. *The teaching gap: Best ideas from the world's teachers for improving education in the classroom*. New York, NY: The Free Press.

Stronge, J. 2002. *Qualities of effective teachers*. Alexandria, VA: ASCD.

Supovitz, J. A., and H. Turner. 2000. The effects of professional development on science teaching practices and classroom culture. *Journal of Research in Science Teaching* 37 (9): 963–980.

Yore, L., J. Anderson, and J. Shymansky. 2005. Sensing the impact of elementary school science reform: A study of stakeholder perceptions of implementation, constructivist strategies, and school-home collaboration. *Journal of Science Teacher Education* 16 (1): 65–88.

Chapter 8

K20: Improving Science Across Oklahoma

Jean Cate, Linda Atkinson, and Janis Slater

Rationale

Since 2005, the University of Oklahoma's K20 Center for Educational and Community Renewal has provided science resources and professional development to over 600 teachers in 54 elementary and middle schools across the state of Oklahoma. In 2005 and again in 2008, the K20 Center was awarded the bid for the three-year, statewide K–8 science professional development institute (PDI) sponsored by the Oklahoma Commission for Teacher Preparation and funded by the Oklahoma legislature. The contract provides $300,000 to work with approximately 100 teachers and administrators each year. The formal goals of the program are to

- deepen teacher content knowledge in science;
- introduce and support research-based teaching practices that increase student conceptual understanding in science; and
- create supportive school contexts that foster teacher and student science learning.

The K20 Center approaches these goals through an integrated whole-school professional development model that emphasizes leadership and the development of professional learning communities. This model is based on the IDEALS (Table 8.1) and 10 research-based practices associated with high-achieving schools (O'Hair, McLaughlin, and Reitzug 2000) (Table 8.2, p. 122). The model has been used extensively and has demonstrated its effectiveness for increasing teacher knowledge, creating professional learning communities, and increasing student achievement (Williams et al. 2008).

Table 8.1 IDEALS for democratic education

Inquiry	Critical study of practice by the school community
Discourse	Substantive conversation focused on teaching and learning
Equity	Seeking fair and just practices for all stakeholders
Authenticity	Learning that is constructed through disciplined inquiry
Leadership	Shared understanding that leads to a common vision
Service	Serving the needs of the greater community

Source: O'Hair, McLaughlin, and Reitzug 2000.

Table 8.2 10 practices of high-achieving schools

Practice 1	Shared vision	**Practice 6**	Inquiry and discourse
Practice 2	Authentic teaching and learning	**Practice 7**	Supportive leadership
Practice 3	Shared leadership	**Practice 8**	Community connections
Practice 4	Personalized learning environments	**Practice 9**	Concern for equity
Practice 5	Teacher collaboration	**Practice 10**	Access to external expertise

Source: O'Hair, McLaughlin, and Reitzug 2000.

The K20 Science PDI targets improving science teaching in prekindergarten through eighth-grade science through whole-school professional development. In year 1, teachers receive inquiry science kits and technology supported by monthly onsite professional development. In year 2, support for implementation continues for new teachers with several site visits. Results show an increase in teacher science knowledge, use of inquiry, and student learning.

Teachers who previously lacked confidence in their teaching of science share with excitement how their students respond to inquiry science.

> I think the kids really enjoy doing [science]. I would see them talking about experiments and science ideas. It changed their attitude toward science too.
>
> Students love science, the messier the better. ... We were racing water drops, and the kids loved it.
>
> I am not the world's best science person, so having someone show me and be there monthly to give me other ideas to expand on was more than I thought I would get. (Quotes from Science PDI teacher participants)

Yet, there is still pressure for these teachers to prioritize math and reading over science.

> Sometimes we can't get to science because math and reading take forever. (Quote from a Science PDI teacher participant)

This chapter will describe how the model has been used by the Oklahoma Science PDI to increase student and teacher learning in science and move schools toward actively integrating research-based teaching methods in their classroom practice.

Model

Background and Context

The Science PDI is one program in the university–school partnership of education projects provided through the K20 Center at the University of Oklahoma. The K20 Center is an educational research facility designed to address the needs of the 21st century learner and workforce development through schoolwide systemic change. Using a five-phase model, the K20 Center provides professional development for (1) effective school leadership strategies;

(2) whole-school faculty learning using technology as a catalyst; (3) teacher content-specific learning to improve student achievement; (4) authentic learning strategies for student engagement; and (5) scholarships and support for undergraduate science, technology, engineering, and mathematics (STEM) students. Authentic learning and teaching create a climate that nurtures the 21st-century skills. By embedding innovative technology into authentic learning and teaching, the K20 Center supports school transformation and renewal. Individual programs within each of the five phases are integrated to provide opportunities for professional growth that ultimately impact student learning and achievement in literacy and STEM content areas.

The gateway program and cornerstone of the K20 Center professional development continuum is the Phase 1 Technology Leadership program. This program introduces school leaders to strategies and networking opportunities for the development of technology leadership focusing on 21st century learning. The principals and/or superintendents of all schools participating in any of the programs offered through the K20 Center's school-university partnership are required to participate in Phase I Leadership. This leadership development program ranked third in the nation in creating leaders for systemic, substantive change to support student learning when evaluated (NSDC 2004). Leaders attend a two-day seminar with follow-up quarterly meetings, ultimately producing an action plan for technology integration in their school that focuses on at least 1 of the 10 research-based practices (O'Hair, McLaughlin, and Reitzug 2000) that are emphasized in the program. The emphasis on leadership is the key to the success and sustainability of programs offered in the other phases of the K20 Center's professional development model (Williams et al. 2008).

Teachers and students benefit significantly from the K20 professional development offered in each of the phases. Examples of research and evaluation findings include the following:

- Technology grant schools funded by the Oklahoma Educational Technology Trust (OETT) and supported by K20 professional development show a greater increase in the State of Oklahoma's Academic Performance Index than the state's average increase (Williams et al. 2008).
- K20 Center's STEM initiatives indicate a significant increase in teacher content knowledge and change in teaching practice, while students whose teachers participate in K20 professional development score significantly higher on science process tests and show increased learning (Greene 2010; McKean and Fredman 2008).
- In the U.S. Department of Education–funded digital game–based learning project that produced *McLarin's Adventures*, which uses the principles of instructional game design to integrate challenging eighth- and ninth-grade core content standards, findings show significant learning gains for students who participated in the digital game–based learning treatment (Atkinson 2009).

In addition, industry-sponsored funding for STEM scholars supports 20 undergraduate students from K20 network schools, promoting STEM learning and increasing opportunities for students to pursue STEM-related fields.

Science PDI Approach to Professional Development

The university–school partnerships formed in the Science PDI through the K20 Center provide sustained professional development, science materials, and technology equipment to schools serving students in prekindergarten through eighth grade. The professional development in these schools targets a deeper understanding of science content and pedagogy through the creation of communities of practice focused on STEM learning. As teams of teachers work together to implement and integrate science strategies that are supported by research and focus on student-centered instruction, learning communities are strengthened, resulting in student success.

Approximately 8–10 schools with a total of 100 teachers are selected for the Science PDI annually through a statewide application process. The application is posted on the K20 Center and Oklahoma Commission for Teacher Preparation websites and an announcement is distributed via email to the 2,000 school leaders who have attended the Phase 1 Leadership seminars. The application asks for a description of the school's efforts toward implementing the IDEALS (see Table 8.1, p. 121), effective science learning, and a professional learning community. Science test scores for the school as a whole are also requested, including a report of the disaggregated scores showing the breakdown for individual learning objectives. The whole-school approach to professional development requires that at least 80% of the teachers who teach science in the school agree to participate and sign the grant application. The school principal must also agree to participate in leadership-specific professional development and meetings during the course of the initial grant year. The program design targets systemic change and relies on the participation of the entire school community. Any public school in the state of Oklahoma may submit a grant application, and if the principal has not participated in the Phase 1 Leadership program, he or she agrees to do so during the grant year.

Schools that are selected to participate in the Science PDI receive Full Option Science System (FOSS) kits, technology equipment, and monthly school site visits by a K20 science specialist. During the monthly visits, the science specialist provides professional development that uses multiple strategies, allowing teachers to learn about the science content of their grade-level kit as well as other content related to the state science standards. The professional development sessions emphasize research-based science teaching practices, technology integration, and integration of science with other content disciplines. The science specialist works with the principals and teachers to create a needs assessment that informs and drives the type of professional development each school receives. Informal formative assessment throughout the year allows for adjustments and additions to the professional development that is provided as new needs are identified.

Each grade level teacher or pair of teachers receives a Delta Education FOSS kit. Kit selection for each grade level of the program is based on its relationship to the state science standards and the overall student performance on state tests. The FOSS program, developed by the Lawrence Hall of Science at the University of California, Berkeley, is designed to engage students in actively constructing science knowledge through structured inquiry.

FOSS kits combine inquiry, hands-on, multisensory, cooperative, interactive, and reflective instructional practices. This foundation supports the goals of the program and aligns with the professional development provided to the schools. The kits provide a practical, structured way to introduce teachers to inquiry instruction and helps teachers experience and understand the impact of research-based teaching practices.

The initial professional development session occurs at the K20 Center in the summer prior to the first year of the grant. Teachers from all of the grant schools attend a daylong session that focuses on the FOSS kit they will receive for their grade level and the associated pedagogy that is used in the kit to engage and teach for conceptual understanding. The administrators also attend a full-day session that orients them to specific grant expectations and provides them with an overview of the teaching philosophies that will be emphasized over the course of the grant in addition to supportive leadership strategies. This hands-on session provides an opportunity for the administrators to experience some of the same kinds of learning that will be provided for their teachers as well as an opportunity to discuss and understand its value for students with other administrators.

After the initial session, most of the professional development takes place at the school site and is tailored to meet the needs of each participating school. This adaptive approach to the design of professional development represents a constructivist view and allows each school to start from wherever they are in the use of inquiry learning practices and progress from there. The K20 Center science specialists (former elementary and middle school teachers) provide follow-up support and coaching through monthly visits to the school, supplemental instructional materials, and online mentoring for the first year. In the first session at the school site, the student performance data of each school is reviewed, and areas of weakness in the objectives are identified. In almost all cases these weaknesses are found to occur in experimental design, identification of variables, and interpretation of data. The professional development needs for the school are assessed using these data as a starting point. Often teachers have never looked at data in this way, and it provides them with an opportunity to express needs for their own professional learning in terms of how it will most profoundly affect their students. Subsequent professional development is structured to address these needs.

The professional development schedule is determined collaboratively within each school to fit the school structure and the preferences of the staff. The science specialists work with school leadership to creatively structure as much time as possible for professional development. Time and substitutes often present a challenge to schools with increasingly tight budgets and demands for state and federal educational mandates. School visits are frequently embedded into the school day, with science specialists meeting with a small group of teachers while substitutes rotate to relieve one small group of teachers, then another. Some schools prefer to meet after school. Each school is asked to provide time for at least two whole- or half-day sessions in order to allow for extended, in-depth learning and whole-staff collaboration. There is money budgeted in the grant to reimburse schools for substitutes in order to offset budget constraints.

Sessions are designed and programmed to provide teachers with support for teaching their FOSS kit, understanding and implementing inquiry learning practices, adapting and adding science lessons that address state science standards, and incorporating strategies that integrate literacy and mathematics with science. The framework that informs the professional development sessions for this program is the student/assessment/content-centered approach for designing classroom environments that is described by Bransford, Brown, and Cocking (2000) in *How People Learn: Brain, Mind, Experience, and School*. More recently, the professional development approach was extended to include formative assessment strategies and learning progressions that address basic concepts required for conceptual understanding. Using this constructivist approach for the design of professional development, we ensure that the basic goal of providing teachers with strategies to increase conceptual understanding for themselves and their students is addressed, yet we start teachers at their own level of prior knowledge and move them forward rather than taking a one-size-fits-all approach to professional learning. In evaluation data, teachers reported that the approach to professional development helped them begin looking at science instruction more holistically and that it increased collaboration between teachers.

In addition to the monthly sessions, the principal and a team of teachers attend the K20 Center's annual conference, the Innovative Learning Institute, as well as an end-of-year reflection and celebration day. Principals also attend a required session during the school year to learn more about building supportive conditions for science education that includes walk-through observations with reflective conversations targeting science instruction. This approach to instructional supervision is adaptable and can be expanded to all content areas by targeting research-based pedagogy and instructional practices that support critical thinking and problem-solving.

A limited amount of technology equipment is provided to each school as an incentive for participation as well as a vehicle for collaboration. The type and amount of technology varies from year to year depending on budget and availability. In general, most schools receive about $3,000 of technology equipment to supplement their science kit award. Professional development provides strategies to integrate the technology into the instructional program. Examples of technologies provided in various years include scientific probeware, interactive slates, electronic balances, document cameras, projectors, netbooks, and audio equipment. Although this technology is not the focus of the grant, it has proven to serve as a motivator for participation and leads to collaboration centered on integrating technology into authentic instructional practices. The first year of the grant is the most intense, although follow-up is provided in year two. New teachers to the school are invited to attend the summer sessions, and two follow-up sessions are provided at each school site during year 2. During the site visits, the science specialist meets with new teachers to help them orient to the science materials and with the whole staff to continue the support to sustain the changes effected in the previous year.

External evaluation results show that the Science PDI is meeting its program goals and is making a difference in the schools that participate in the program. Evaluation data show an increase in teachers' science learning, confidence, and inquiry usage. School leaders from PDI schools share that teachers are implementing more inquiry and technology in their science teaching and that students are participating in engaging, hands-on activities. Furthermore, science kit module assessment results show that elementary and middle school students exhibit statistically significant growth in knowledge in the science concepts presented, and the percent of students scoring proficient on the state science assessments increases significantly (McKean and Fredman 2008).

Schoolwide Implementation Examples

> My feeling toward science instruction definitely has changed. It was a dread, and I really didn't want to do it because I was used to doing the units. But with the hands-on, I have just as much fun as the kids, and to be honest, I have actually learned how to branch out. (Quote from a Science PDI teacher participant)

The philosophical underpinnings of the Science PDI relate to the philosophy of the K20 Center. The K20 Center partners with schools as they adopt the IDEALS systemic change model and 10 research-based practices associated with high-achieving schools (O'Hair, McLaughlin, and Reitzug 2000). Examples of teacher implementation of the IDEALS through the Science PDI are shown in Table 8.3. Substantive systemic school change progresses through shared leadership, ongoing reform-style professional development, technology as a catalyst for change, and school–university partnerships.

Table 8.3 IDEALS in action in the Science PDI schools

Inquiry	A middle school science department uses professional development time to construct a common definition of inquiry learning in order to create a vertically articulated set of process skills to improve their scores in experimental design on state assessments.
Discourse	Fourth- and fifth-grade teachers from two schools in a small but growing district meet during monthly professional development sessions to collaborate on common benchmarks for each grade level in order establish districtwide consistency and promote collaboration.
Equity	An early childhood center applies for the science PDI grant in order to make sure that all students receive science instruction in the early grades after discovering that science instruction is inconsistent from classroom to classroom.
Authenticity	An elementary resource teacher leads students in an interdisciplinary investigation of local water quality. The students are invited to present their findings at the University Research Day conference for graduate and undergraduate students.
Leadership	A school district with five elementary schools and no science curriculum director creates a science advisory board consisting of teacher representatives from each school in order to improve science instruction across the district.
Service	A middle school science staff partners with a local electrical utility to have students identify and quantify energy use in the school in order to make recommendations for energy-saving ideas to help with school budget cuts.

Source: O'Hair, McLaughlin, and Reitzug 2000.

A look at two elementary schools within the K20 network viewed as successful in implementing the Science PDI grant will provide insight into the effectiveness of this statewide university–school network professional development approach. It will also illustrate how schools in the program follow individual paths to success through the adaptive nature of the professional development while maintaining fidelity to the authentic inquiry practices supported by the Science PDI. Both elementary schools are involved in the K20 Center's statewide network of schools, with both of the leaders previously participating in the K20 Phase 1 Leadership program. The leaders are active in the network, and both schools ultimately received an OETT technology grant for $40,000 worth of technology equipment and yearlong intensive professional development from the K20 Center. School 1 received the Science PDI grant a year before the technology grant, whereas school 2 received the technology grant a year after receipt of the Science PDI grant.

School 1 is typical of many of the K20 Network schools. It is a rural school located in a remote area of the state. The school has one administrator and 10 classroom teachers who participated in the Science PDI. The students in this prekindergarten through eighth grade school are 58% Caucasian and 42% Hispanic, with 62% qualifying for free or reduced lunch subsidies. Changes in demographics have occurred over the past several years for this school. In 2000, the student population was 78% Caucasian and 22% Hispanic, with 46 % of the students on free or reduced lunch subsidies. The percent of first through third graders receiving reading remediation services is 30%. The attendance rate at parent conferences is 99%. For 2008, the school enrollment was 400. The science specialist's field notes describe school 1 as follows:

> During the grant year, the school made good progress toward switching their focus to hands-on, inquiry learning. This very enthusiastic staff was actively involved in getting the grant and committed themselves to completely integrating the kits into their ongoing instruction. They designated an unused classroom as the science "lab" for those activities from the kit that required extra space or equipment. The lab was busy all year. Teachers commented that they think much more now about their teaching practice, and they expressed pride in their efforts to raise their state test scores, which went up from last year. They are looking to continue the trend.

School 1 state science test scores show 80% proficiency in 2008, with 36% scoring advanced, whereas in 2009 test scores show 96% proficiency, with 41% scoring advanced.

School 2 is a rural school located near a large urban district and several suburban districts. Two administrators and 18 teachers participated actively in the Science PDI. The student demographics show 77% Caucasian, 3% African American, 1% Asian, 2% Hispanic, and 17% Native American, with 37% qualifying for free or reduced lunch subsidies. The percent of first through third graders receiving reading remediation services is 42%. The attendance rate at parent conferences is 97%. For 2008, the school enrollment was 463. School 2 is described in the science specialist's field notes as follows:

> Teachers utilized the FOSS kits in their class successfully, though they seemed to consider them as a separate science curriculum. The staff was excited about beginning to implement some form of hands-on science in their classrooms. They struggled to overcome obstacles to teaching science and found that the student engagement was worth the time and effort. Teachers talked about finding ways to increase their science resources in the future so that their students would have more opportunities to experience science outside the textbook "units" that they had been using in the past. They were very eager to find ways to draw on science contexts to work with their students on reading and math to help improve their test scores.

School 2 test scores show 83% proficiency in science in 2008, with 28% scoring advanced, whereas in 2009, test scores show 85% proficiency, with 28% scoring advanced.

In general, the value of the following conditions are apparent in these successful PDI schools: (1) leadership that includes supportive conditions, collaboration, and a shared vision for science learning; (2) a school–university partnership that encourages systemic change and provides external expertise for supporting increases in science content and pedagogical knowledge; and (3) technology integration that provides additional support for implementation of reform-based science instruction.

Lessons Learned

Most science and math teachers work in isolation, with little support for innovation and few incentives to improve their practice. Yet much of teachers' best learning occurs when they examine their teaching practices with colleagues. Research (Putnam and Borko 2000; Printy 2008) indicates that teachers are better able to help their students learn science and mathematics when they have opportunities to work together to improve their practice, time for personal reflection, and strong support from colleagues and other qualified professionals. Desimone (2009) proposes five core features of professional development that can change teachers' practice. These core features include content focus, active learning, coherence, duration, and collective participation that lead to increased teacher knowledge and skills, which change teachers' beliefs and, in turn, advance the changes in instructional practices that promote improved student learning.

School 1 and school 2 are examples that epitomize the multisession, sustained professional development model used by the Science PDI, which extends professional learning over time to provide teachers the opportunity to collaboratively examine teaching practice and reflect on its value for their students. Systemic changes that occur in these Science PDI schools can be explained by the conditions nurtured within the program design, supportive leadership, support for increasing content and pedagogical knowledge through the university–school partnership, and efforts toward technology integration for authentic learning.

Leadership

As schools strive to implement content-specific reform for improved student learning, leadership does make a difference. When leaders provide a common vision and supportive organizational structures and build the capacity of teachers, student learning increases (Leithwood and Wahlstrom 2008). Professional community is a mediating factor, while the level of school support serves as an enabling condition in the development of teacher knowledge and change in practice (Desimone 2009; Ingvarson, Meiers, and Beavis 2005). When implementing new strategies, principal and teacher encouragement enhances the process (Printy and Marks 2006). Effective instructional leadership is a key to school improvement and helps teachers develop their professional capacity. This is the rationale for the inclusion of administrators in all aspects of the Science PDI, including administrator-specific professional development sessions, and also for requiring that they participate in the Phase 1 Leadership program.

Supportive Conditions Increase Openness to Implementation

In successful PDI schools, administrative support is apparent. Teachers from both schools discussed above indicate that they feel supported by their administrator in their efforts to implement the science kits and strategies introduced in the initiative. A typical comment is, "[Our principal] is always there for us when we have questions or need help. He helps us with resources and materials as much as he can." In turn, the administrators indicate that they fully support science instruction because it is "good for the students." Administrators from both schools talk about being out in their teachers' rooms on a regular basis and engaging them in discussions about what is going on in their classrooms.

Support was less obvious at the district level in each of these schools. Neither school district has a science curriculum supervisor or a specific budget for science instructional materials. School 2's district has vertical alignment sessions in all content areas, but most teachers are unaware of the outcome of the science alignment. In both schools, teachers indicate that reading and math are the focus of both testing and curricular decisions, and that science is of secondary importance. In a sense, the K20 specialists function as temporary, external curriculum specialists for these schools. The school administrators, while acknowledging support for science instruction, also indicate that their teachers have a first responsibility to reading and math. The solution they see as most feasible to the problem is integration. A typical administrator comment is, "The science grant gave me a new perspective on how to help the teachers incorporate science with the reading and extended vocabulary."

Collaboration Supports the Change Process

The perception of the administrators in both schools is that more collaboration is occurring in their school but not in a structured setting as evidenced by the comment, "Teachers that I didn't think were even talking to each other from the upper and lower grades are talking about ideas they have shared." The school 2 administrators share that they perceive the Science PDI grant as bringing in an element of collaboration and teamwork to the staff that had been missing. "On top of the team planning that occurred with the K20 specialist, vertical planning also happened because they were talking about the science and what was going to happen the next year and what happened the year before. We want to continue this and make sure it is going on, and it has as far as we know." Groups of teachers also seem to be coming together to implement the science kits and activities, especially within grade levels. As an example, third-grade teachers at school 2 devised a structured plan for teaching their kits. "We worked together. Each of us took a part of the lesson and we rotated the kids. I felt like it flowed, and it was easy the way we worked it out."

A Shared Vision for Science Learning Guides Implementation

Both teachers and administrators indicate that their schools have "no real vision" for science instruction as a separate entity, although one teacher states that "it is really something we should do." The school 1 administrator refers specifically to authentic learning and its association with science inquiry as a direct link to the whole-school vision. Both teachers and administrators refer to science pedagogy most often in terms of the "hands-on" approach as compared with traditional textbook instruction. Yet, the idea of aligning instruction with research-based practices is still noticeably missing from the conversation.

In both schools, traditional teacher-led instruction with the textbook as the curriculum guide appears to be the norm as stated by both teachers and administrators. It is indicated that, in accordance with the school vision, movement away from this type of pedagogy is desired but not always possible. The main barriers seem to be the extra preparation and instructional time needed for activities and space in which to do them. "We would probably like to do more hands-on if we had the option, but in reality, how do we get through it? You know we have such a push for math and reading and, honestly, that's what we are tested on. In a perfect classroom, you would like to have the hands-on, but ... you know." Their comments support the prevailing school attitude that science is not as important as other subjects. One teacher from school 2 exhibits conflicting attributions for the significance of science instruction when she says, "I know that science is really important but it is so challenging to try to squeeze it in." The teachers in school 1 appeared to have less trouble doing science in their classes, though they still articulated challenges of time and space. One teacher says, "I try to do more science now instead of leaving it out or doing it later, realizing that it doesn't have to be a book to be science." Another goes even farther in saying, "We don't seem to need textbooks as much anymore. We see it now as a place for information, a resource rather than a curriculum." The emphasis on authentic teaching and learning in this school's vision may account for the difference seen here in contrast to the other school's vision for integration into reading and math.

Collaborative Leadership Supports the Struggle to Overcome Barriers

Despite the barriers to implementation, teachers overall indicate a shift in their personal feeling toward science instruction. Descriptions such as, "I learned to trust that if I could step back, the kids could do it (inquiry learning);" "the more you realize how much the hands-on stuff helps the kids, the more you try to pull it in wherever you can;" and "I'm not sure if their learning has been impacted, but the kids loved it" are common. Their comments show that, although they may not yet see the pedagogical impact of an authentic inquiry approach to instruction, they see student engagement increase. More evidence of the lack of science emphasis when compared to math or reading is illustrated by this statement: "We did our kit in the spring because it was a good way to pass the time away. We did it in the afternoon during testing because the kids were stressed, and we didn't want to focus on academic stuff as much or worry that something might get left out." We see this struggle for science time in schools as the biggest challenge for science improvement. As science comes more and more to the forefront in educational conversations at the state and national level, this attitude is gradually changing. We feel that the Science PDI provides opportunities for substantive conversation within school communities that contributes to the gradual shift in attitude and practice that is emerging in the current climate of reform lead by the initiative to implement the *Common Core State Standards* and the *Next Generation Science Standards*. Our PDI schools will be able to approach the change proactively rather than in a reactive way.

School–University Partnership

Schools that have access to external sources of support show success (Allen and Hensley 2005; Lieberman and Miller 2007). Through external support for school reform, university partners create professional communities among schools and interest groups across schools (Lieberman and Miller 2007). By increasing professional learning across schools, participants develop enthusiasm about teaching and learning (Hargreaves and Goodson 2005). School–university partnerships facilitate changes in teaching and increase student learning (Atkinson et al. 2009).

The school–university connection in the form of the professional development relationship with the K20 Center appears to facilitate change. A teacher from school 2 says, "Having somebody show you how to go about doing the kits and answer your questions helped tremendously. I knew that our science specialist would be coming every month, and she was wonderful." A school 1 teacher comments, "The grant motivated me to bring in science more than I had been. The personal visits were very helpful in finding ways to put stories and science together." A school 2 administrator noted that, "The follow-up with the science specialist was the big key. Otherwise, I don't know if we would have had buy-in as well as we did."

Schools that participate in the Science PDI seek to incorporate science learning into their school vision but often do not know how to accomplish it. The external knowledge provided by the K20 Center, which is focused on ideas grounded in current research, helps schools to find ways to change their current views of teaching and learning that might not otherwise be accessible to them. As teachers attempt to learn new pedagogies and look for new ways of thinking about student-centered learning and instruction, they begin to form a community of practice that influences the whole-school vision in innovative ways that bring in the critical thinking and problem-solving approaches associated with science learning.

Technology Integration

Technology can serve as a catalyst for teacher learning (Williams et al. 2008). Key elements for the role of technology in professional development include (1) meeting the learning needs of the participants, (2) having skilled facilitators, (3) connecting content with teachers' practice, (4) developing and facilitating a learning community, and (5) establishing mechanisms for reflection (Loucks-Horsley et al. 2003). Educational technology use can impact student achievement, self-concept, and quality of interactions between teachers and students (Bialo and Sivin-Kachala 1996).

A limited amount of technology equipment was provided to both schools through the Science PDI grant. Both of these successful PDI schools are also associated with the K20 Center through the OETT grant that provides technology to Oklahoma schools. Interestingly, the principals often credit technology as a key component and are eager to describe the effect technology has had on their instructional practice and staff motivation.

School 1 received a technology integration grant prior to the Science PDI. The principal speaks about the impact that the technology received through these grants has had on his school. He talks about an increase in the level of collaboration within the school and how his teachers are continually finding new and authentic ways to use technology in the classroom. He stresses, however, that the most important thing about the technology has been the "new way of looking at instruction." He indicates that the professional development provided by the K20 Center helps teachers become more collaborative as they work to integrate it into their practice in an authentic, productive way rather than seeing it as "playing with toys." He also cites a strong relationship built on trust with the K20 staff during the sustained professional development process as helping the teachers decide to pursue the science grant and to effectively integrate the innovation into their school culture. It is his feeling that receiving the Science PDI grant after the technology grant is a key factor in moving the teachers toward a more authentic view of teaching and learning, yet, is quick to admit that they still have a long way to go, though he feels they are moving in the right direction.

School 2 received the OETT technology grant and the associated whole-staff professional development for technology integration two years after their Science PDI grant, and the administrators credit the Science PDI grant with helping to change the culture of their school to a more collaborative one. They feel that the sustained professional development nurtured this culture and helped inspire their teachers to make a school decision to apply for the technology grant. They admit that technology was the driving force all along because they had determined this to be a need for their school. The science grant was originally seen as a means to that end, but they feel they "got much more from it than they had expected." It helped the staff think of technology as more than a skill-drill tool or a presentation device but more as a way to improve instructional practice and engage students. The administrators are excited about the enthusiasm and collaboration the new technology has generated.

Reflections for the Future

It is interesting to note that, although the avenues to the science grant are different for both schools, both have cited collaboration and external support as key factors responsible for their growth. A noteworthy difference lies in the fact that the school that had the technology grant first seems more focused on authentic teaching and changing practice, whereas the school that has not yet experienced the technology grant is focused on the motivational and supplementary instructional value of the technology. The emphasis on authenticity in the professional development that supports the technology grant may yet again move their view of science instruction beyond merely fitting in science or pulling it into reading and math to a more authentic integrated view.

Leadership is critical to teacher reform initiatives. Its importance cannot be overstated. Leadership includes district administrators, school leaders, and teacher leaders, with each providing valuable support. Those in leadership require ongoing professional learning and communication as they build productive communities that impact student learning. When leaders create supportive conditions, the implementation of inquiry science and technology is enhanced. Supportive conditions include the availability of professional development structures, modeling, trust, and resources. The ability of leaders to provide supportive conditions is enhanced through the school–university partnership and through the additional resources provided by other grant projects. Supportive conditions in which collaboration is focused on instruction changes practices. Teachers who focus on learning about science inquiry share formally and informally about classroom practices, thus stimulating implementation. In some schools, teachers serve as science leaders and mentors, enhancing the implementation of science. Having a shared vision for science-specific instruction encourages the implementation of teaching reforms. When the reform teaching strategies are embedded within a culture that embraces the reform-style instruction, the science reform is accelerated and sustained.

The implementation of inquiry science teaching and learning practices introduced through the sustained professional development process impacts the school context through coherence and collective learning. The school–university network supports the learning and leadership for principals. The network provides teachers with outside support and encouragement for making changes in practice that they see as important. Developing teacher capacity with ongoing learning encourages reflection on teaching practices that leads to implementation gains. Professional development that is sustained, personal, and focused, yet adaptive to the school context and the needs of the teachers, facilitates substantive and lasting change. The Science PDI reflects successful professional development impacting teacher learning and student learning through content focus, active learning, coherence, duration, and collective participation (Desimone 2009). Inquiry science provides the content focus and the active learning components. Teachers prepare and teach inquiry-based lessons with the monthly support of the K20 professional development mentor. The inquiry-based lessons fit into the larger context of authentic teaching and learning, providing a coherent view of learning. This coherence provides widespread support for the content reform and is supported through the leader in the broader context. Evidence affirms the need for more than one year of support for changes in teacher practice. The technology grant provides one year of intensive support, while the science grant provides another year of intensive support with a second year of follow-up support.

A focus on collaboration allows for collective participation and moves reform-type teaching strategies through the school community. Teachers find support in knowing that others are working on similar strategies and are able to discuss their struggles and successes with colleagues. Teachers check their own understanding with colleagues and encourage each other to implement the teaching of science. Focusing on whole-school change, rather than trying to change practice through individual teacher workshops, can increase the momentum of change in practice.

Conclusion

Professional development that focuses on shared vision and shared leadership does make a difference. Leaders who continue learning, who bring innovative practices to their schools, and who bring external expertise to assist with their reform efforts, can bring about changes in teacher practice and student learning. In addition, changing to inquiry science takes time, resources, and focus. University–school partnerships provide the needed support for the change process. When leaders and teachers work toward a common vision with external support, change occurs. Within supportive conditions for collaboration and a vision for science learning, implementation of reform science instruction is possible. Technology integration encourages collaboration as teachers learn new strategies for teaching and learning with the implementation of new technologies. It seems that teachers are willing to admit they need help in implementing technology and that has some carryover to their accepting assistance in other areas of teaching and learning. With relationships of trust built from consistent support for authentic teaching and learning from the school–university partnership, teachers are open to implementing research-based practices that increase science learning and achievement for their students.

References

Allen, L., and F. Hensley. 2005. School–university networks that improve student learning: Lessons from the league of professional schools. In *Network learning for educational change,* ed. W. Veugelers and M. J. O'Hair, 17–32. New York: Open University Press.

Atkinson, L. 2009. *STAR: Networks for emerging technology schools (NETS).* USDE Award U203G050016. Norman, OK: University of Oklahoma.

Atkinson, L., J. M. Cate, M. J. O'Hair, and J. Slater. 2009. K20 model: Creating networks, professional learning communities, and communities of practice that increase science learning. In *Professional learning communities in science: Lessons from research and practice,* ed. S. Mundry, and K. E. Stiles, 129–148. Arlington, VA: NSTA Press.

Bass, H., Z. Usiskin, and G. Burrill, eds. 2002. Studying classroom teaching as a medium for professional development: Proceedings of a U.S.–Japan workshop. Washington, DC: National Academies Press.

Bialo, E. R., and J. Sivin-Kachala. 1996. The effectiveness of technology in schools: A summary of recent research. SLMQ 25 (1).

Bransford, J. D., A. L. Brown, and R. R. Cocking, eds. 2000. *How people learn: Brain, mind, experience, and school*. Expanded edition. Washington, DC: National Academies Press.

Desimone, L. M. 2009. Improving impact studies of teachers' professional development: Toward better conceptualizations and measures. *Educational Researcher* 38 (3): 181–199.

Greene, B. 2010. *K20 research experiences for science teachers (KREST)*. NSF REESE Grant 0634070. Norman, OK: University of Oklahoma.

Hargreaves, A., and I. Goodson. 2005. Series editors' preface. In *Network Learning for Educational Change*, ed. W. Veugelers and M. J. O'Hair, vii–ix. Maidenhead, England: Open University Press.

Ingvarson, L., M. Meiers, and A. Beavis. 2005. Factors affecting the impact of professional development programs on teachers' knowledge, practice, student outcomes, and efficacy. *Education Policy Analysis Archives* 13 (10): 1–28.

Leithwood, K. K., and K. Wahlstrom. 2008. Linking leadership to student learning: Introduction. *Educational Administration Quarterly* 44 (4): 455–457.

Lieberman, A., and L. Miller. 2007. Transforming professional development: Understanding and organizing learning communities. In *The keys to effective schools: Educational reform as continuous improvement*, ed. W. D. Hawley, 99–116. Thousand Oaks, CA: Corwin Press.

Loucks-Horsley, S., N. Love, K. E. Stiles, S. Mundry, and. P. W. Hewson. 2003. *Designing professional development for teachers of science and mathematics*. 2nd ed. Thousand Oaks, CA: Corwin Press.

McKean, K., and T. Fredman. 2008. *K20 science professional development institute: Year three implementation and outcomes evaluation report*. Cushing, OK: Oklahoma Technical Assistance Center.

National Commission on Mathematics and Science Teaching for the 21st Century. 2000. *Before it's too late: A report to the nation from the national commission on mathematics and science teaching for the 21st century*. Washington, DC: U.S. Department of Education.

National Science Teachers Association (NSTA). 2004. Position statement: Scientific inquiry. *www.nsta.org/about/positions/inquiry.aspx*.

National Staff Development Council (NSDC). 2004. *Building for success: State challenge grants for leadership development. Report of a study*. Seattle, WA: Bill and Melinda Gates Foundation.

Newmann, F. M., M. B. King, and P. Youngs. 2000. Professional development that addresses school capacity: Lessons from urban elementary schools. *American Journal of Education* 108 (4): 259–299.

O'Hair, M. J., J. J. McLaughlin, and E. C. Reitzug. 2000. *Foundations of democratic education*. Fort Worth, TX: Harcourt College Publishers.

Printy, S. M. 2008. Leadership for teacher learning: A community of practice perspective. *Educational Administration Quarterly* 44 (2): 187–226.

Printy, S. M., and H. M. Marks. 2006. Shared leadership for teacher and student learning. *Theory Into Practice* 45 (2): 125–132.

Putnam, R., and H. Borko. 2000. What do new views of knowledge and thinking have to say about research on teacher learning? *Educational Researcher* 29 (1): 4–15.

Williams, L. A., L. C. Atkinson, J. M. Cate, and M. J. O'Hair. 2008. Mutual support between learning community development and technology integration: Impact on school practices and student achievement. *Theory into Practice Journal* 47 (4): 294–302.

Chapter 9

The iQUEST Professional Development Model

Katherine Hayden, Youwen Ouyang, and Nancy Taylor

This chapter provides insight into the investigations for Quality Understanding and Engagement for Students and Teachers (iQUEST) model by describing the iQUEST approach and key components of the professional development, discussing the impact and fidelity of the iQUEST model on participating teachers and students, and sharing lessons learned and future directions of the model.

Rationale

The iQUEST model was developed through a project funded by the Innovative Technology Experiences for Students and Teachers program at the National Science Foundation (NSF) and evolved from a collaboration between educational technology and computer science professors at California State University San Marcos (CSUSM) and the K–12 science coordinator for the San Diego County Office of Education. As such, the project brings technology-enhanced learning experiences as an early intervention to middle school students into classrooms having high percentages of traditionally underserved populations in science, technology, engineering, and mathematics (STEM) fields.

One cohort of seventh-grade and one cohort of eighth-grade science teachers participated in iQUEST professional development and mentoring activities for two years, during which time the teachers received support in becoming part of a cyber-ready workforce who felt confident integrating technology into classroom activities to enhance student understanding of science concepts. The project was designed under five guiding principles:

1. Students' best chance to experience technology-enhanced learning comes from lessons planned by teachers who are confident in using technology.
2. Students and teachers increase 21st-century workforce skills through technology-enhanced learning experiences.
3. Students' and teachers' individual needs are addressed in learning communities.
4. Students who are engaged in hands-on investigations have deeper understanding of science concepts.
5. Students who see themselves as scientists pursue STEM careers.

The project leadership team believes that digital resources such as visualization tools, interactive simulations, Web 2.0 tools, videoconferencing, and social networking and collaboration applications promote student interest in technology and science and prepare students to be technologically literate for the 21st century workforce. The iQUEST strategy is to get teachers to become knowledgeable and more comfortable with using technology tools and the integration of these tools into everyday science classroom experiences. The iQUEST two-year professional development model provides professional development to teacher teams during a summer academy by allowing team members to identify and align science and technology resources with grade-level curriculum and content standards. Through Collaborative Lesson Study (CLS) rotations during the school year, project teacher teams develop, test, and refine lessons that target gaps in student learning through integration of technology tools and resources within classroom activities that optimize the learning of science content. This team approach has proven to be effective in increasing teacher confidence in a collaborative, supportive way.

The iQUEST project promotes putting technology in the hands of students to support teaching and learning, which can require modification of existing technology policies and allocation of resources. For instance, some schools had not allowed use of Web 2.0 tools or provided students with the ability to download files such as KMZ files that are designed to enhance students' Google Earth experiences. The iQUEST leadership facilitated discussions between teachers and the district technology department to bring about changes that allowed students the ability to download files and in other cases, access website collaborative tools that were previously blocked. As teachers identify free online resources for classroom activities, support from information technology administration is also critical to ensure the timely update of systems and network setup. To secure the buy-in of district administration and school principals, the iQUEST leadership team invites representation of administrators to its advisory board and continuously showcases teachers' successes and evidence of student learning to provide insight into the value of the professional development and innovative technology uses embedded in teacher lessons.

Model

The iQUEST professional development model was designed to increase teacher confidence and to expand application of technology in support of teaching and learning of science. From 2007 to 2013, the model has been refined based on its implementation for teachers in 11 public school districts, one charter school, and one private school in Southern California. More than 75 teachers, mostly from schools with high populations of diverse learners, have participated in professional development using this model within three NSF-funded projects at CSUSM. The School of Education at CSUSM has a longstanding partnership with local school districts through their site-based clinical practice programs for credentialing teachers. This allows the project leadership to work closely with district administrators to identify and recruit science teachers. In some cases, teachers sign up in grade-level teams, anxious to codevelop their curricula; in other cases, individual teachers sign up to increase their ability to collaborate with like-minded teachers from other schools and districts as they pioneer new technologies and pedagogy for teaching science.

Teachers were not required to have specific technology skills prior to participating, but were asked to commit to two years in the project. During the first year, teachers develop foundation skills and pilot introductory lessons using technology. They expand and reinforce their skills during the second year, and lessons become more advanced in use of technology and more student-centered.

In planning the professional development for teachers, the leadership incorporated Knowles principles from adult learning theory to make the experiences meaningful and effective for teachers (TEAL 2011). There are several assumptions in adult learning (referred to as andragogy) that are different from the pedagogy used for working with K–12 students, and these were important to address. These assumptions include understanding that adults (1) need to know why they need to learn; (2) want acknowledgement that they are capable of self-direction; (3) have accumulated more experiences than younger learners and need to have their prior experiences valued; (4) are motivated to apply learning to immediate and real-life situations; and (5) are responsive to motivators such as stipends, advancement, certifications, and so on (Knowles, Holton, and Swanson 2005; TEAL 2011).

It is also critical for adult learning experiences to include a set of procedures that involve the learner in the process of preparing, establishing climate, diagnosing needs, formulating program objectives, designing and conducting the pattern of learning experiences, and evaluating the experiences (TEAL 2011). With these foundational assumptions, the iQUEST professional development model was designed to include summer academies, collaborative lesson study rotations, and monthly afterschool workshops. Throughout the school year, teachers have access to follow-up support, mentoring, and a collaborative online system to allow teachers to engage in sharing and reflection on lesson implementation.

Summer Academies

The summer academies provide opportunities for teachers to center their learning around student needs, challenge misconceptions, acquire deeper content understanding, and think deeply about how they have come to understand scientific concepts that can be supported through technology. Building on decades of research on how adults learn (TEAL 2011), professional development activities reinforce how teachers can integrate and transfer what they learn to their classrooms by applying new ideas and technology resources to lessons they see as needing modification to increase student understanding. As such, the iQUEST summer academies scaffold learning experiences on to existing curriculum used in classrooms through the use of interactive simulations and visualization tools.

The PhET Interactive Simulations project at the University of Colorado at Boulder, which was founded by Nobel Laureate Carl Wieman, offers individual exploratory environments in which formulas and theories are put into real-life contexts. Figure 9.1 (p. 140) captures a group of middle school–level simulations from the PhET website (*phet.colorado.edu*). These simulations make the invisible visible, for instance, by showing electrons, photons, or molecules. They allow experiments with equipment that is normally not available in middle school settings. In addition, students can explore just the idea being tested without having to take too many real-world complications into account.

Figure 9.1

MIDDLE SCHOOL–LEVEL PHET INTERACTIVE SIMULATIONS

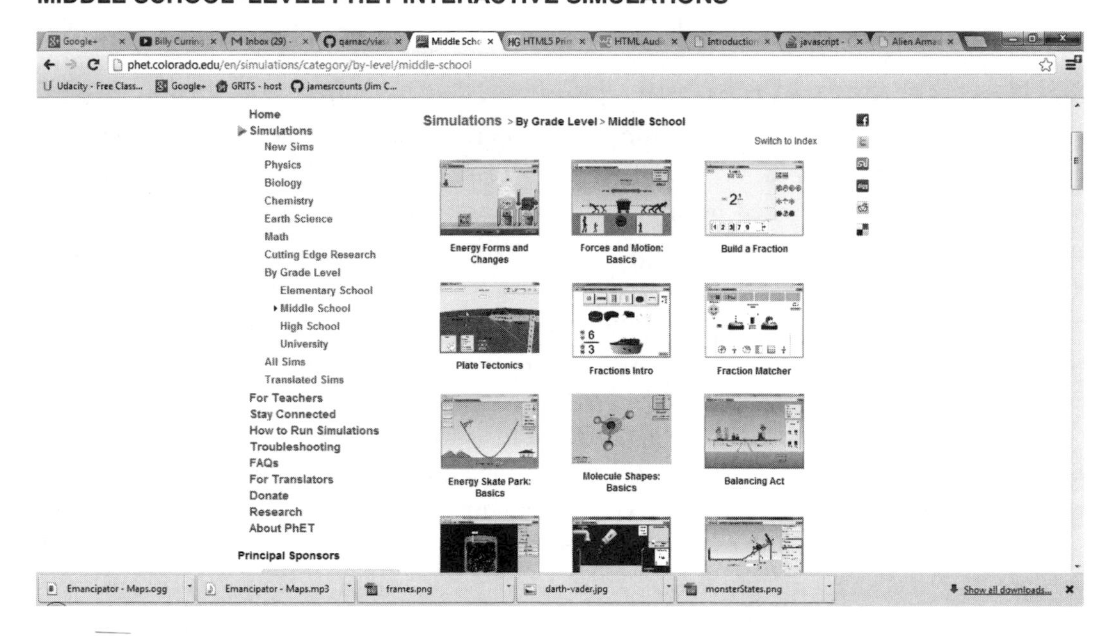

Forming teams, teachers identify tools of their choice with which to experiment, discuss potential misunderstandings about science concepts, and then plan strategies for effective integration of the tools in alignment with student learning goals and standards. For example, through the Moving Man simulation from PhET (Figure 9.2), teachers design a lesson that helps students see the relationship between motion and its resulting graph. Teachers' own experimentation with the simulation helps them identify probing questions for students, such as, "What do negative values in the graph tell you about the man's movement?" "Why is the graph moving when the man stands still?" and "Given a story of the man's activities, how do you generate a chart to reflect his movement?"

The academies also introduce Web 2.0 tools that support student collaboration. Teachers explore a variety of these tools from a list of recommendations provided by the project (*sites.google.com/site/khaydenweb20*). Many of these Web 2.0 tools provide students with an opportunity to explore science concepts or present their knowledge and understanding of concepts. One example tool is Popplet (*popplet.com*), which allows students to work collaboratively to create a concept map showing connections between concept components using text and images. The resulting maps can be shared with a larger audience online.

The iQUEST summer academies also provide teachers with opportunities to interact with scientists. Science topics are introduced by experts who present information that

Figure 9.2

POSITION VS. TIME CHART FROM THE PHET MOVING MAN SIMULATION

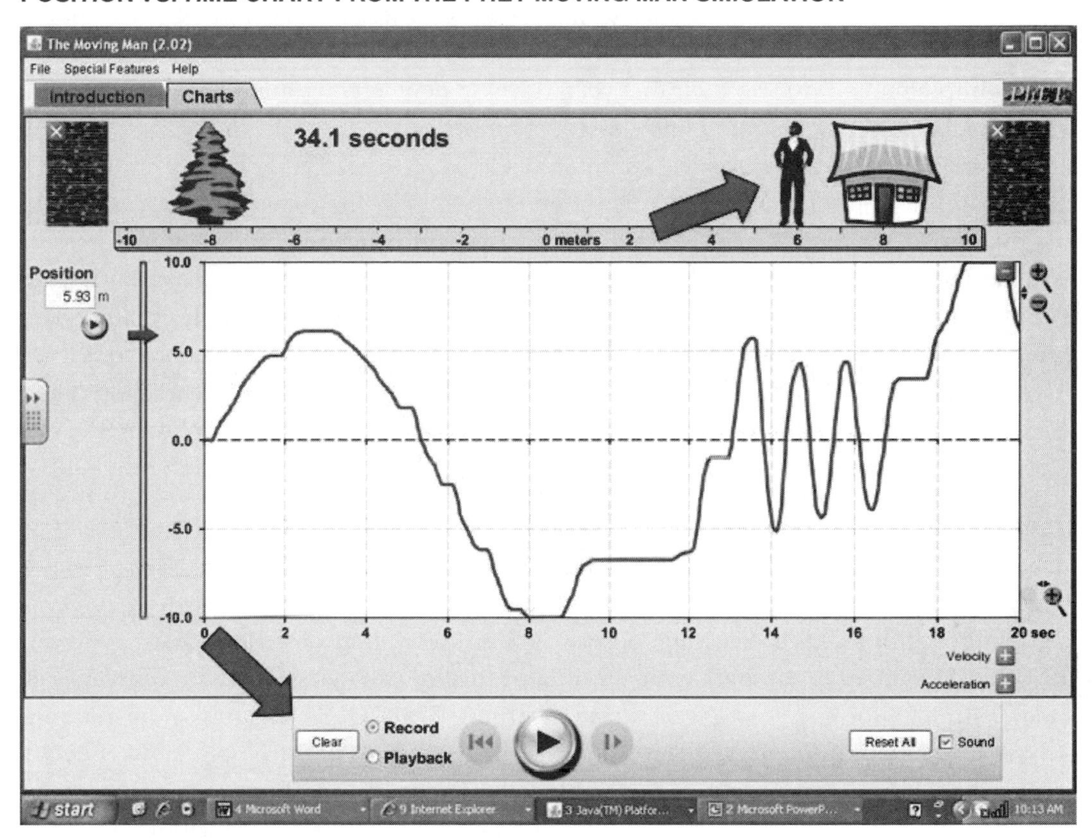

expands teacher knowledge and connects real-world science with the textbook-driven science curriculum. Scientists can be brought into the academy virtually, in person, or through a field trip. For instance, teachers took a field trip to the Beckman Institute for Conservation Research at the San Diego Zoo Safari Park where they interacted with scientists at the institute to learn about current research projects. During the field experience, the director of genetics at the institute gave a tour of the Frozen Zoo and explained how technologies have supported molecular genetic analysis, including automated DNA sequencing and other advanced methods for studying genetic variation. Through videoconferencing, Dr. Maureen Ferran, an associate professor of biological sciences at the Rochester Institute of Technology, presented her research on viruses and their connection with science standards. She discussed appropriate online resources with teachers after assessing their scientific accuracy and relevancy to life science standards. One web application was designed to help students understand cloning. The teachers were so intrigued with their virtual meeting experience that some of them invited Dr. Ferran to videoconference with their classes to answer student questions after their exploration of the Click and Clone Web application.

The second-year summer academy builds on the collaborative community created in year 1 as well as on iQUEST lessons created by project teachers during their first year. Teachers are offered the opportunity to go deeper into lesson development and reflect on the impact of technology uses in their classroom. Working in teams, teachers refine lessons implemented during their first year, develop new lessons, and identify additional resources that can be integrated into classroom activities.

As a culminating event for the summer academy, teachers complete an individualized action plan in which they identify strategic points in their curriculum at which to integrate project activities and lessons. They also provide information describing areas in which they need additional support during mentoring times and select workshops they will attend during the coming school year to further develop their technology skills. The monthly workshops offered by the project are selected by surveying teachers about their interests and needs. Teachers are required to attend five of the monthly workshops each year; however, many attend more. Three example workshops are Digital Probes in Science, Global Positioning Systems (GPS), and Google Earth (see Workshop section for details).

CLS

In a school culture based on clear standards, information about how students actually learn science is essential to instructional improvement (Donovan and Bransford 2005). A key component of CLS is teaching, observing, and debriefing of lessons that have been cocreated by a team of teachers with a facilitator to allow analysis of student outcomes as a major strategy for improving teaching and learning. Thus, CLS promotes understanding of how students respond and understand content, leading to the development of effective science teaching (Perry, Lewis, and Akiba 2002). The real lesson from this collaborative process is not the product but the process. The design of iQUEST's CLS includes elements from the BSCS 5E Instructional Model (Bybee et al. 2006), research on how people learn (Donovan and Bransford 2005), principles of adult learning theory (TEAL 2011), and use of professional communities of practice that help teachers to stay connected and share outcomes with their peers.

Teachers participating in iQUEST engage in four CLS experiences over their two years of professional development. Each CLS experience includes a full day for planning, a day of team-teaching rotations and collaborative reflection and refinement of the lesson at a school site, and a final debrief day during which teams polish the lesson for publication and share the lesson with other teacher teams. The next section describes the CLS process used by iQUEST as part of its professional development model.

CLS Planning Day

iQUEST project leaders, teachers, and facilitators convene on a Saturday to initiate the lesson development process. Working in groups of three to four teachers at the same grade level, teachers meet with a trained facilitator to coplan their lesson. The role of the facilitator is essential to assure a risk-free environment during all aspects of the CLS, setting norms for collaboration, maintaining a focus on the science concepts developed at each stage of the 5E Model planning, ensuring careful consideration when selecting appropriate technology tools or resources, and engaging all members of the team. The facilitator leads the team in a backwards design process (Wiggins and McTighe 2005). Before learning activities are brainstormed or designed, the team is charged with articulating what their students will be able to explain about the science concept at the end of the lesson. Beginning with the intended outcome, the student's explanation of the science concept is a purposeful component that contributes to development of learning activities that best support student learning of the topic. This is the most challenging piece of CLS because teachers typically prefer to talk about what *they* will do, when in fact a focus on what students do to engage in deeper understanding of a concept is the learning goal for CLS. The facilitator coaches the CLS team through the 5E Model with a foremost lens on what the student does and a secondary look at what the teacher does. The 5E Model lessons are scripted electronically using a collaborative online tool such as a wiki or Google site and Google Docs. All of the logistics to deliver the lesson are defined, shared, and available online for every team member to access.

Another significant outcome of the CLS planning is the professional dialog about the science concept the teachers have identified as one with which students typically struggle. As teachers negotiate how to scaffold instruction to address the concept, there are multiple opportunities for teachers to redefine their own understanding and learn the content more deeply through discussion with their peers. The identification of technology resources that might assist students in developing a deeper or clearer understanding of a topic further challenges science teachers as they search to identify technology resources for the lessons. One group of teachers adopted the use of digital microscope probeware to help students understand that all living organisms are composed of cells and that transportation of materials is an essential function of the cell membrane. By connecting digital microscopes (see Figure 9.3, p. 144) to regular classroom microscopes, the magnification level was greatly enhanced to allow students to see the detailed parts of cells. These digital microscopes are also connected to laptops via USB, allowing digital images to be captured and shared within groups and allowing the whole class to better collaborate and engage in meaningful discussions.

CLS Team-Teach Rotations

CLS rotations allow teachers to try out their lessons in a real class session setting. Each CLS team convenes at a host school prior to the start of the school day to review the plan for instruction and confirm the logistics of materials, online access, and the schedule for alternating between teaching and observing versus debriefing and refining the lesson. A typical schedule includes three designated class periods or sections for instruction with time for debriefing and refining the lesson. Two teachers teach the first lesson as scripted while other team members observe and take notes on student responses to the lesson. Observers are asked not to interact with students during the lesson in order to maintain an authentic learning environment. After the first teaching of the lesson, the CLS team gathers student work and meets to debrief. The norms of collaboration are reviewed to ensure that the intent of the debrief is to focus on the student outcomes and not the teacher. The facilitator oversees a discussion culminating in reflection on the entire process (i.e., sharing of perceptions about the content and student engagement, timing of the 5E Model plan, and the resulting student products). Then, modifications are agreed on by the team, and they prepare for the second team-teach rotation followed by another debrief and reflection time for fine-tuning of the lesson. Student artifacts provide evidence of the effectiveness of the modifications made to the lesson during each iteration of the rotations.

CLS Collaborative Debrief

During the collaborative debrief phase of CLS, all team members convene and complete a protocol incorporating the review of student work while determining how the technology components of the lesson support the learning objectives and how the resulting student work may indicate learning gaps in the content, the process, or the communication of the science concepts. This step in the CLS process helps teachers in understanding more clearly the connection between the learning opportunity and the student outcomes. Although the main objective of CLS is the process of practicing the 5E Model lesson protocol and articulating appropriate lesson strategies that support content acquisition, the collaborative debrief process allows refinement of CLS lessons to be shared with a broader audience. The results of the refined lessons representing Earth, life, and physical science content at the middle school level are available on the iQUEST website (*www.csusm.edu/iquest*).

Figure 9.3

USE OF DIGITAL MICROSCOPE TO SUPPORT INVESTIGATIONS AND COLLABORATION

Elodea Leaf - Plant Cell

1,000 Magnification

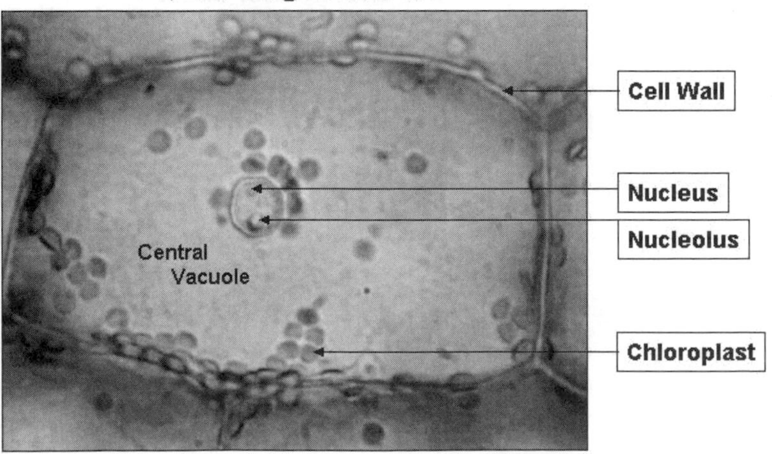

Support, Mentoring, and Workshops

Another critical component of the iQUEST professional development model is the ongoing support for teachers as they implement technology-enhanced lessons in their classrooms. The project coordinator serves as a mentor during the summer academy and through monthly site visits and virtual communications with teachers. Computer science and science education students support classroom implementation of multiple lessons each semester. They work with teachers to set up and test technology ahead of time and then assist by providing technical support during the lessons.

Monthly afterschool workshops provide teachers with additional technology and science content training while allowing teachers to meet face-to-face and discuss how well lessons worked (or if they didn't).

The project also introduced teachers to the use of GPS receivers for science-related activities by engaging them in geocaching activities. A high school science teacher led the GPS workshop and provided a rich discussion to elicit connections to science concepts. Teachers were placed in teams and gained firsthand experience and understanding of GPS receivers. The workshop provided teachers with opportunities to brainstorm ideas for the use of GPS receivers in science lessons.

The iQUEST model keeps teachers connected to their peers who are simultaneously involved in the professional development experience. In a project in which teachers are located at different sites or regions, technology acts as a bridge to connect the professional learning community. The project course management system features social networking tools that engage teachers in online professional dialog through chats, wikis, blogs, forums, journals, and other tools. These experiences, which are intended to support professional dialog during the school year, also assist teachers in adapting teaching strategies to use technologies that are attractive and familiar to digital-age students (Prensky 2001). Teachers are provided the opportunity to request support for developing a class page where they post links and lessons for their students.

Lessons Learned

Assessing Project Impact

Through observations of classroom teaching and the CLS process, the external evaluator of the iQUEST project documented changes in teaching over time. Summary findings for the CLS evaluation include the following:

- The three-part CLS process provides adequate time for teachers to thoroughly explore, implement, and refine technology-based science activities for students. Small groups of four teachers were useful for encouraging input from a variety of voices that addressed characteristics of different schools and students, helping make lessons useful across a broad teacher audience.

- The iterative process of teach-discuss-refine in the teaching rotations is critical to teachers being able to improve their lessons. Although teachers often improve lessons with practice, the peer feedback opportunities associated with CLS enabled rapid improvement in lesson delivery. A full debrief day devoted to reflection, refinement, and documentation of lessons results in polished lessons, full of helpful hints for integrating technology tools that could be used by other teachers. This process was necessarily accompanied by rich discussions about effective pedagogy as teachers discovered and tried new methods of facilitating learning.
- The CLS process also offers teachers a chance to build a professional network that is based in learning content-relevant technology skills and lesson development. Increased skills with technology tools and their integration into science curriculum has led some teachers to procure equipment, such as digital microscopes and probes, financed by their schools.

Project teachers complete a survey, both at the beginning and the end of each school year, reflecting via Likert-scale questions (1 "strongly disagree" to 4 "strongly agree") their use of technology tools, confidence with using those tools, number of iQUEST lessons taught, beliefs about the use of educational technology, satisfaction with program components, how the program affected them, and the degree to which the program emphasized student interest in STEM. Table 9.1 suggests that teachers felt iQUEST had a positive impact on their ability to consider, use, and advocate for technology in their teaching. Given the relatively high ratings of some items both before and after the school year, the statistically significant gain for advocacy is especially noteworthy. iQUEST professional development—CLS in particular—is oriented toward classroom implementation by teachers.

The 2011–2012 iQUEST student survey collects data about outcomes associated with STEM interest and technology use, as well as demographic data. The survey was completed by seventh-grade students from 15 iQUEST teachers in 11 San Diego area schools during both fall 2011 and spring 2012. The analysis of survey results shows changes in scores over time, with iQUEST students demonstrating significant gains in technology use, confidence using technology, and interest in computing careers. The project also assessed seventh-grade content gain on life science topics, including DNA, basic cell structure and function, basic plant and animal biology, and other related topics. The final, cleaned data set contained 1,908 observations, or pre- and postscores for 954 students from 13 teachers. From fall 2011 to spring 2012, gains on average scores were significant, with an effect size (pre to post) of 0.633.

Dissemination to Districts

The iQUEST project works with teachers to develop leadership skills in lesson development and presenting during dissemination workshops. Through the collaborative process of planning, implementation, debrief, and refinement, teachers document exemplary lessons in iQUEST modules using a template that includes

- an overview of the lesson;
- science concepts and learning objectives addressed by the lesson;
- activities and concepts prior to the lesson;
- supporting technology resources required for the lesson;
- detailed lesson elements described in the 5E Model;
- recommendation for assessment of student learning; and
- teachers' notes that capture experiences gained from the field-testing of the lesson in their own classrooms.

Table 9.1 Summary of teacher survey on professional development impact

Impact	*n*	Fall 2011		Spring 2012		Change from 2011 to 2012
		Mean	SD	Mean	SD	
I often think about appropriate technology for a lesson	15	3.80	0.41	3.87	0.35	0.07
I advocate for technology to support science learning	15	3.53	0.52	3.87	0.35	0.33
I now have more access to technology in the classroom	15	3.67	0.49	3.73	0.46	0.07
I seek ways to acquire more technology resources (grants, site funds, etc.)	15	3.33	0.82	3.33	0.49	0.00
I mentor other teachers in planning and using technology for student learning	15	3.13	0.74	3.27	0.59	0.13
I feel comfortable presenting technology at conferences	15	2.87	0.52	3.07	0.46	0.20
School or district administrators ask me to provide leadership to others	15	2.73	0.70	3.00	0.65	0.27
I will continue to use iQUEST lessons in coming years	15	NA	NA	3.93	0.26	NA
I will continue to identify ways to provide students with information on STEM careers	15	NA	NA	3.87	0.35	NA

The project actively seeks opportunities for teachers to share exemplary lessons and classroom experiences through presentations and workshops at local, regional, and national teacher conferences, including the annual conferences held by Computer-Using Educators, the California League of Middle Schools, the National Science Teachers Association, and the International Society for Technology in Education.

Through mentoring from the project leadership, teachers develop more confidence in presenting and showcasing their successes. As a result, they are ready to become leaders and assist in the adoption of iQUEST professional development in their home districts. To disseminate iQUEST professional development beyond its partner districts, the project leadership works with regional county offices of education to host all-day workshops for teachers and administrators. During these workshops, iQUEST teachers present exemplary lessons and engage workshop teachers in identifying ways to adapt and adopt iQUEST lessons for their own classrooms. These workshops open the doors for excitement and interest in STEM learning. Some schools that are involved in the project have used the CLS model for site professional development activities.

Implications for Use in Implementation of the Next Generation Science Standards

The iQUEST model of professional development aligns with the new *Common Core State Standards* (NGAC and CCSSO 2010) as well as the *Next Generation Science Standards* (*NGSS*; NGSS Lead States 2013). With the *NGSS* focus on the essential importance of training in STEM, emphasis on exploration and inquiry, and connections with real-world applications, professional development is needed to prepare teachers to expand the use of technology and incorporate these essential components into their curriculum. iQUEST engages teachers in a collaborative process that models these components within the professional development environment and relies upon 21st-century online tools to support the process.

The *NGSS* are built on three dimensions including practices, crosscutting concepts, and core ideas. This aligns well with the focus in the iQUEST teacher academy of providing connections with scientists, scientific organizations, and current research practices in order to engage teachers in deepening their learning in science content and expanding their connections within the scientific world. The crosscutting dimension links the different domains of science. The iQUEST model provides the opportunity for teacher teams to build lessons collaboratively and test and refine them through CLS rotations. CLS opportunities for teachers will lead to strong lesson development using the crosscutting framework.

iQUEST incorporates understanding and investigating as an element of the lesson study process for teachers. In addition, iQUEST focuses on increased interest in science and technology for students and provides teachers the opportunity for working together in grade-level teams while participating and sharing within a multigrade project that allows discussion across grade levels.

Reflections for the Future

The iQUEST Leadership Team works closely with the project's advisory panels to ensure the work is moving toward all project goals and continues to refine and document the professional development model's success. They have identified essential elements that are important to these endeavors based on the lessons learned during implementation.

The first essential element is to lay a solid foundation in the project goals and team building. The initial teacher academy is scheduled during summer when teachers are relaxed and do not have pressures of the school year. A framework of five days was found to work best, as opposed to the three-day summer academy first initiated. Five days are needed in order to ensure that teachers have a solid foundation in the project goals; time to boost teacher skills and knowledge in both science and technology; and time for team building, collaboration, and lesson development. The project also began with an orientation prior to summer so that teachers were prepared for the summer academy and familiar with the overall project outcomes and expectations.

The second essential element of the iQUEST professional development is to have trained facilitators who become the key to success in each stage of the CLS process. These facilitators must be knowledgeable in science and technology and have skills in managing teacher team dynamics. The leadership team trains facilitators ahead of time and provides a facilitator handbook on the iQUEST CLS process.

Finally, the third essential element of iQUEST is ongoing mentoring and support for teachers. This includes working directly with teachers monthly and connecting with district leadership (such as the technology administrators) to support teachers moving through the implementation of the project lessons. The key to successful mentoring is to customize the support to the individual needs of each project teacher and plan strategies that meet the needs of busy teachers with compact schedules. Flexibility and patience are important to keep teachers connected when other stressful situations demand their attention.

The leadership team is currently implementing the iQUEST model of professional development in two additional NSF projects on the basis of the success of the model in prior projects. One of the current projects is multistate and expands the iQUEST model to sustain implementation in schools and districts through dissemination efforts.

References

Bybee, R. W., J. A. Taylor, A. Gardner, V. P. Scotter, J. C. Powell, A. Westbrook, and N. Landes. 2006. *The BSCS 5E Instructional Model: Origins and effectiveness*. Colorado Springs, CO: BSCS.

Bybee, R. W., J. A. Taylor, A. Gardner, V. P. Scotter, J. C. Powell, A. Westbrook, and N. Landes. 2006. *The BSCS 5E Instructional Model: Origins, effectiveness, and applications: Executive summary*. Colorado Springs, CO: BSCS.

Donovan, M. S., and J. D. Bransford, eds. 2005. *How students learn: Science in the classroom*. Washington, DC: National Academies Press.

Knowles, M. S., E. F. Holton, and R. A. Swanson. 2005. *The adult learner: The definitive classic in adult education and human resource development*. 6th ed. Burlington, MA: Elsevier.

Lewis, C. 2002. What are the essential elements of lesson study? *California Science Project Connection* 2 (6).

National Governors Association Center for Best Practices and Council of Chief State School Officers (NGAC and CCSSO). 2010. *Common core state standards*. Washington, DC: NGAC and CCSSO.

NGSS Lead States. 2013. *Next Generation Science Standards: For states, by states*. Washington, DC: National Academies Press. *www.nextgenscience.org/next-generation-science-standards*.

Perry, R., C. Lewis, and M. Akiba. 2002. Lesson study in the San Mateo-Foster City School District. Paper presented at the annual meeting of the American Educational Research Association, New Orleans, LA.

Prensky, M. 2001. Digital natives and digital immigrants. *On the Horizon* 9 (5): 1–6.

The Teaching Excellence in Adult Literacy (TEAL) Center. 2011. TEAL center fact sheet No. 11: Adult learning theories teaching excellence in adult learning. TEAL. *teal.ed.gov/tealGuide/adultlearning*

Wiggins, G., and J. McTighe. 2005. *Understanding by design*. Alexandria, VA: ASCD.

Chapter 10

The Boston Science Initiative: Focus on Science

Sidney Smith and Marilyn Decker

Rationale

The success of an urban-based science initiative in Boston Public Schools from 2001 to 2008 required the confluence of many variables and partners that all too often do not cohere within a large, urban district. The success in part can be attributed to a nationwide, urban-based systemic initiative funded by the National Science Foundation (NSF). Boston Public Schools competed for and received funds to help improve the teaching and learning of science at the K–12 level. At that time in the district, Boston Public Schools was deep in the reform of mathematics and English language arts, but science was not on the agenda. Fortunately for science supporters in the district, the goals and objectives of the urban-based initiative closely aligned with the then superintendent's passionate focus on teaching and learning. This endorsement by the superintendent helped to catalyze the science movement and resulted in extraordinary changes for years to come (Yin 2006). This chapter, then, focuses on science as part of a larger ongoing districtwide reform effort as seen through the lens of professional development.

Model

When science came to be included in the district's reform plan, this step called for changes that were virtually unimaginable in the Boston Public Schools of the time. The district had little experience with horizontally and vertically aligned curriculum in any subject area, let alone science. Existing policies allowed many students to graduate from high school with no more than eighth-grade competence in mathematics, which, in turn, limited the number of students who had any chance of completing rigorous high school science courses. Professional development, if any existed at all from school to school, was haphazard, unfocused, and left to the strategic thinking of principals and their instructional directors. Local stakeholders within and outside the district (e.g., colleges and other educational institutions) often drove professional development efforts that were aligned with projects being pursued on the respective campuses.

Additionally, funding available for instructional materials and professional development was severely lacking, and no multiyear plan existed to acquire materials and conduct professional development in a strategic and thoughtful manner. The net effect of this approach was that another generation of students (most often disadvantaged and underrepresented) was at a level of academic performance that would not allow them to successfully compete for admission to four-year colleges, competitive technical schools, or demanding jobs. This situation changed as the district geared up to meet the expectations embedded in the inclusion of science in the reform initiative and the Boston Science Initiative: Focus on Science, took hold.

A Change in Policy

Since the level of science teaching at the K–12 grades demanded by the reform movement was new to Boston Public Schools, policies were needed to set the process in motion. By the close of the 2006–2007 school year, the district had adopted new policies that met most of the goals expressed in the science initiative to ensure that the inclusion of science would be more than a passing fancy (see Table 10.1).

Table 10.1 Sample elementary teacher activity log

Mathematics	
Grade	Time on task
K–5	70 minutes/day
6	80 minutes/day
7–8	45 minutes/day
9	80 minutes/day
10–12	45 minutes/day

Science	
Grade	Time on task
K–2	90 minutes/week
3	135 minutes/week
4–8	225 minutes/week
9–12	225 minutes/week

Overall, new policies were put in place that

- increased (significantly) the time students would devote to science for teachers of grades K–12;
- required all high school students to take and pass three laboratory courses in science;
- required elementary science specialists to devote a minimum of 12 hours to professional development per year, funded and provided by the district;
- aligned the standards in all grades with the state curriculum frameworks;
- aligned all science curricula within the district both horizontally and vertically;
- created a funding schedule (and districtwide Science Materials Center) that ensured consumable science materials would be replaced on an annual basis and wholesale materials adoptions would take place every five years;
- required middle and high school science teachers to devote 24 hours to professional development in science per year, funded and provided by the district;
- required each school to include explicit goals, objectives, and benchmarks in science, as well as related professional development plans, in its annual Whole-School Improvement Plan; and

- generated a system of annual, mid-, and end-of-year assessments (and related reports) in science for teachers of grades 6–12.

Again, as was true with the inclusion of science teaching in K–12, these policies, and the rapidity with which they were developed and implemented, was unprecedented in Boston. However, as anyone engaged in education knows, goals, objectives and policies are often as solid and binding as the paper they are written on. It would take far more than paper to turn these major policy mandates into measurable changes in teacher and student performance.

Leadership Development

As much as anything, the success of the science inclusion relied on a committed group of strategically placed district, school, and classroom leaders who worked very long hours to ensure the success of the project. After being given a green light (and minimal structure) by the superintendent, an array of leaders designed and pursued a top-down and bottom-up strategy that would dramatically change the face of science instruction in the district.

From the top, the director of curriculum and instructional practices translated reform efforts into policy proposals and managed the approval process by generating support from the superintendent and his leadership team and the Office of Technology regarding the implementation, scoring, and dissemination of data associated with the new district-wide assessment program. Critical to the development of tools for change, the initiative also involved the senior program director of science (and other subject-area directors) and a staff member in the Science Office. As important as anything, during this process, top leadership officials initiated curriculum implementation reviews across the district for mathematics, history, English language arts, and science.

These reviews were assessed by senior program directors in actual teachers' classrooms in as many schools as possible each year through visits beyond the regular performance evaluation process. The senior program directors also assessed the implementation of the science (and other) programs in each school, including administrative support of the programs using a program-assessment rubric. The resulting data and explicit improvement plans were presented to and discussed with the principal of each school, as well as the deputy superintendents who evaluated the principals.

These steps produced even more confidence in the reform effort from the superintendent and assured him that the initiative was serving as a significant vehicle for change in teaching and learning. As a consequence, the district was able to secure millions of dollars in additional funding from multiple sources over the life of the reform to help sustain the effort.

Building Science Department Leadership Capacity

Concurrent with the top-down movement, working from the bottom up, the senior program director of science and a very small staff worked diligently to build a cadre of elementary, middle, and high school teacher leaders across the district (see Table 10.2, p. 154).

Table 10.2 Teacher leaders by grade band and demographics

Demographic	Elementary	Middle	High School	Total
Male	10	13	12	35
Female	16	10	18	44
Asian	3	1	1	5
African American	4	10	9	23
Hispanic	1	0	2	3
White	18	12	18	48
Total	—	—	—	79

While the director of curriculum and instructional practices generated a secure funding stream that provided schools with high-quality materials on an annual basis, the senior program director crafted a top professional development program that would ensure those materials were put to the best use, resulting in the level of professional development delivery depicted in Table 10.3.

Table 10.3 Sample professional development activity and log by a teacher leader

School	# of teachers	Workshop hours	Teacher leader log: average # of hours	Total
Elementary	26	49	67	116
Middle	23	49	79	128
High	30	49	76	125

At the start of Focus on Science, the leadership team consisted of five people who had never worked closely together—a newly named science director, three professional development specialists, and one materials management specialist. Four of the team members had been in Boston Public Schools for many years as classroom teachers, department chairs, and science office support personnel. The newly named science director had extensive experience leading major projects but only a cursory understanding of the culture of the Boston Public Schools. To get the project going, the team needed time to work together to develop

- a coherent vision for what a good science classroom looks like;
- a strong understanding of the Massachusetts science standards;
- a good understanding of exemplary curriculum materials;
- strategies for selecting, piloting, and implementing curricula;
- skills in designing and delivering professional development;

- skills in facilitating group discussions; and
- an understanding of how to manage materials.

With limited time and mostly external funding, the work had to begin immediately and to show progress from the beginning (Yin 2006).

Keeping the Focus and Getting Help From the Local and National Experts

The team turned to local and national leaders for knowledge, help, and inspiration in developing a high-quality professional development program for all K–12 teachers in the district. Within six months of the initiative's inception, science department leaders had developed a coherent vision of good science instruction, revised the district's science standards, and put in place curriculum pilots for grades K–12 (see Table 10.4).

Table 10.4 Instructional materials identification and piloting, grades 1–8

Elementary school pilot materials	
Grade 1	Organisms (STC)
Grade 2	Pebbles, sand, or silt (FOSS)
Grade 3	Changes (STC)
Grade 4	Animal studies (STC)
Grade 5	Ecosystems (STC)
Middle school pilot materials	
Grade 6	Human body systems (STC)
Grade 7	Catastrophic events (STC)
Grade 8	Properties of matter (STC)

FOSS = Full Option Science System; STC = Science and Technology Concepts Program.

These leaders learned early on that before adopting a full set of K–12 curriculum materials and instituting a train-the-trainers approach for professional development, they themselves needed to become skilled in content knowledge and instructional strategies as providers of professional development. To obtain more expertise in curriculum reviews and selections and as providers of professional development, the leadership team attended national and local conferences by, among others, the National Science Teachers Association, the National Academy for Curriculum Leadership at Biological Science Curriculum Study, Merck Institute Science Education Instructional Team Retreat, and the Physics First Symposium at Massachusetts Institute of Technology and Cambridge Public Schools. They also sought support from Curriculum and Implementation Centers, including the IMPACT New England at the Center for Excellence in Science and Mathematics Education, the K–12 Science Center at Education Development Center, and the Smithsonian's National Science Resources Center: Leadership Assistance in Science Education Reform (LASER; *www.ssec.si.edu*).

Through assistance from these programs, the district began an aggressive program of selecting, piloting, and implementing high-quality science curriculum materials. In particular, the Curriculum Implementation Centers provided professional development for the leadership team, information on materials adoption by other urban districts, and guidance on how to structure the adoption process. The actual curriculum adoption plan occurred over a three-year period for all K–12 grades and included a range of products from *Science and Technology for Children, Full Option Science Program, Active Physics, Biology: A Human Approach*, and *Living by Chemistry*.

Focus on Science dramatically influenced and encouraged support for science from universities and other local and national partners. More importantly, relationships with academic partners have continued to improve over time. Focus on Science was instrumental in the direct recruitment of multiple partners that provided a wealth of services and support to the district (see Table 10.5).

The science leadership team included a component in its overall design to keep parents and the broader community continuously informed. Given the new focus on science, information was included in outreach efforts about the state assessment, family science activities, and workshops provided by the Title I Parent Resource Center and communitywide. During the awareness sessions and workshops, emphasis was placed on modeling the type of science being implemented in classrooms and after school activities so that parents and the community had a better understanding of what was happening in the schools.

Teacher Leaders at the Center: Building for Sustainability

The Boston area, like many urban settings, is rich with educational resources and people who want to help. But if the help takes the team in a direction that it does not want to go, it will dilute efforts and undermine the project. Therefore, in the first six months the leadership team worked with the Boston Higher Education Partnership and the Boston Cultural Partnership to coordinate teacher professional development, grant writing, and student programs. In order for any program to be endorsed by the district's science department, it had to focus on the district's students and curriculum.

From the very start of Focus on Science, the leadership team knew that the success of the program would depend on creating a strong culture of teacher leadership (Lord, Cress, and Miller 2008). Coaches for English language arts and mathematics funded by the district were already in place. With full knowledge that science would never have this luxury and that the responsibility for success of any reform would rest with the teachers first and foremost (Davis 2002; Guskey and Yoon 2009; Metz 2008), the leadership team served as the initial providers of professional development.

The development of a strong cadre of well-trained science teacher leaders began in the first year and continues to this day. Although a few district-level science specialists were part of the mix, science teacher leaders were the primary source for science teacher professional development (Fiarman et al. 2009). The focus of the professional development for teacher leaders themselves included immersion in inquiry, curriculum implementation,

Table 10.5 Exemplary long-term partner support

Partners	Contribution
Center for the Advancement of Science and Mathematics at Northeastern University (IMPACT)	Served as a professional development provider for science teacher leaders; hosted and mentored students for science fairs at the school and district levels
Museum of Science, The Children's Museum, New England Aquarium	Provided professional development for science teachers, mentored teacher leaders; provided space for professional development activities; wrote proposals to secure funds for additional professional development institutes and related science activities and materials
Boston College, Lesley University, New England Aquarium, Olin College, Simmons College, Merrimack College, Zoo New England, Massachusetts Institute of Technology, Massachusetts Water Resource Agency, Audubon, Northeastern University, University of Massachusetts–Boston, Boston University	Provided coaching and mentoring of teacher leaders; research opportunities for teachers and students; professional development support for classroom teachers; team leadership development
Massachusetts Institute of Technology, Northeastern University, Harvard University	Provided research experiences for teachers via NSF funding
Boston University	Implemented a state grant to support physics teachers and to help them obtain physics licensure
Harvard Medical School Office of Diversity and Faculty Development	Implemented a grant from the National Institutes of Health to support AP Biology teachers and students
Northeastern University, University of Massachusetts–Boston, Boston University	Implemented the GK–12 Fellows project with fellows attending curriculum-centered professional development through planning with the district's science department
Northeastern University SEED program	Implemented professional development to support the district's science curriculum
Boston Public Schools, University of Massachusetts–Boston, Northeastern University	Collaborated to develop MSP proposal, curriculum implementation, leadership development, and professional development
Northeastern College of Engineering, TechBoston Academy, Tufts Engineering, Concord Consortium, Urban Ecology Institute	Implemented Innovative Technology Experiences for Students and Teachers (ITEST) robotics project

looking at student work, coaching and mentoring, study groups, and case study discussions. The leadership team met monthly with all of the teacher leaders to strengthen the vision of the initiative and provide the teacher leaders with the skills and knowledge needed to work with classroom teachers (Lord, Cress, and Miller 2008).

Initially, the cadre of teacher leaders was made up of 79 teachers recruited districtwide through an application process. Schools whose teachers were selected for the leadership cadre agreed to release these teachers on professional development days. These schools also included an explicit plan for science instruction in their Whole-School Improvement Plans and provided a release period for those teachers the year after they were accepted into the leadership cadre at the middle and high school levels. This cadre of teacher leaders was initially expected to lead professional development within their local schools by leading school-based study groups and to serve as liaisons to the district-level leadership team. However, as the teacher leaders program progressed, the number grew from 79 to 125.

Nearly two-thirds of these 125 teacher leaders participated in more than 115 contact-hours of professional development that covered in-class coaching, assessment, study groups, planning, and piloting on new curriculum material and Looking at Student Work sessions. The goal of the professional development at this point was to provide transformative professional development to the teacher leaders. With this goal in mind, the leadership team designed activities in which teacher leaders could examine their own practice in light of the new task before them, the new policies, and new curriculum materials. The teacher leaders studied examples of good teaching and then applied what they learned in their classrooms. Work by Loucks-Horsley et al. (2003) guided the type of activities the district wanted to see implemented systemwide.

Once in-class professional development got underway, each teacher leader kept a log of their activities (see Table 10.1 for an example of logged activities). Each year, the teacher leaders were given more and more responsibility, and the focus of the work was placed squarely on their shoulders (Lord, Cress, and Miller 2008; Yin 2006). Beyond reviewing, selecting, and piloting new curriculum materials during the first year, the teacher leaders began to provide professional development for classroom teachers as well (Figure 10.1).

As the inclusion of science moved into the second year, new tasks were added to the teacher leaders' lists of responsibilities. For example, they were asked to support other teachers in their buildings during sessions to look at student work and in curriculum implementation. By the third year, teacher leaders took on the responsibility of designing and delivering 12 summer professional development workshops, leading monthly curriculum support sessions, and continuing to provide in-school support for their peers. All the while, they participated in districtwide study and assessment groups. In year 4, the full responsibility for supporting classroom teachers was given to the teacher leaders. In this role, they designed and delivered 22 summer workshops, led in-school support sessions, mentored new teachers, led monthly curriculum support sessions and, for the first time, designed and delivered in-school professional development for their colleagues. The number of classroom teachers involved in the project grew to over one thousand.

By year 5 of the science initiative, the science teacher leaders took on even greater responsibilities (Yin 2006). In that year, the district was awarded a Math and Science Partnership (MSP) grant from NSF. A key part of the grant was the development of graduate-level

Figure 10.1

ELEMENTARY TEACHER LEADER ACTIVITY LOG

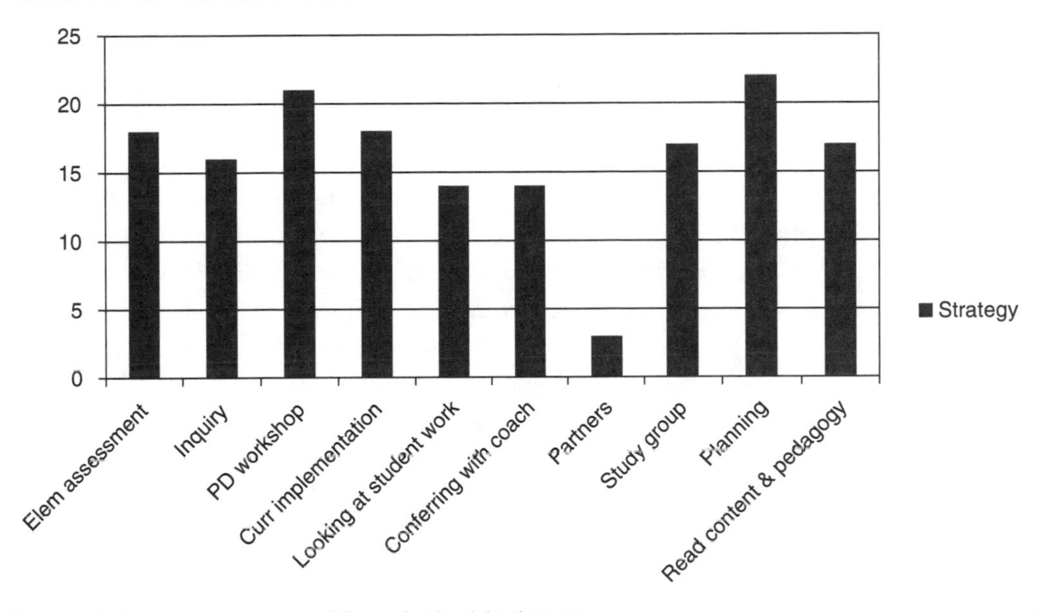

Curr = curriculum; Elem = elementary; PD = professional development.

Source: Boston Public Schools

content courses cotaught by university professors and classroom teachers. By this time, the teacher leaders were experienced workshop leaders and worked side by side with their university colleagues to design and deliver graduate-level courses. The teacher leaders also took on the responsibility of leading vertical teams in their schools and of leading collaborative coaching and learning sessions. These practices continue to this day and would not have been possible without the strong foundation laid by the early work with the urban initiative as just discussed (Fields et al. 2012).

As would be the pride of any district in the country, teachers competed for and earned the Presidential Awards for Excellence in Mathematics and Science Teaching in 2002, 2004, 2007, 2008, and 2010. To come full circle with awareness and training, 139 principals, headmasters, and science supervisors participated in two-day science institutes to help them work with teachers as education leaders. This step bodes well for the need for principals to be highly effective in their roles as educational leaders (USED 2010).

Accountability System

Progress on any significant school initiative is never made without the inclusion of an accountability system that clearly and effectively measures and reports on district, school, teacher, grade level, course, classroom, and student progress (or the lack thereof) over time. The districtwide assessment program in science and curriculum implementation reviews combined to provide a powerful incentive for teachers and administrators to change their practice in the best interests of kids. Individual teacher and school assessment results were distributed districtwide, with every teacher and administrator in the city knowing exactly where each teacher and school stood in relation to all other teachers and schools throughout the district.

The assessment results and the results of the curriculum implementation reviews were shared with the superintendent and deputy superintendents in semiannual debriefings and through individual meetings with the director of curriculum and instructional practices and the senior program director. Data meetings were conducted in schools (some more than others, depending on the influence each deputy superintendent brought to bear on his or her schools) to map out specific strategies to improve teachers' and schools' scores. An unprecedented wave of accountability descended on the schools. Significant accountability tools that could make a difference were now in place.

The director of curriculum and instructional practices, in partnership with the senior program director of science, presented and championed the assessments, the curriculum implementation reviews, and the effective implementation of the districtwide curriculum. These leaders also monitored adherence to the district's policy on instructional time for science and to the implementation of other supportive policies within the district impacting principals, school-based administrators, and teachers. This was done through monthly meetings with these stakeholders and through meetings in individual schools with each school's leadership team. Maybe most important to the success of these presentations was the understanding that this work was a product of over 25 years of relationship building by an individual who was a highly respected teacher, high school headmaster, and central office administrator.

Impact on Student Performance and Achievement

As is true with any reform effort (Spillane and Callahan 2000; Spillane 2012), the intent is to improve teacher performance leading to improved student achievement in science (Duschl, Schweingruber, and Shouse 2007; PCAST 2012). The Boston Science Initiative, Focus on Science, documented an increase in student achievement in science. From 2003 to 2007, students' passing rate on the grade 5 statewide assessment in science improved by nearly 10 points (58–68%), and students scoring at proficient or advanced increased by 4 points (17–21%) (Figure 10.2).

From 2003 to 2007, students' passing rate on the grade 8 statewide assessment in science improved by nine points (37–46%) as shown in Figure 10.3. The number of students passing upper-level high school science courses also increased dramatically.

In 2001, 2,100 students were enrolled in upper-level science courses; in 2007, the number increased to 3,700. Growth in advanced placement (AP) science participation was even more dramatic. In 2000, only 183 students were enrolled in AP science; by 2007, the number increased to 609 students (see Figure 10.4).

Figure 10.2

FIFTH-GRADE SCIENCE STATE ASSESSMENT: PERCENTAGE PASSING, BY RACE/ETHNICITY

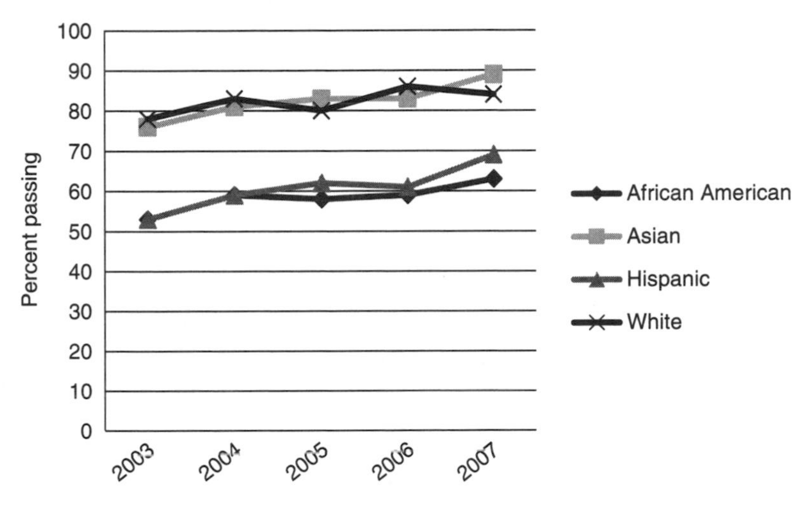

Source: Boston Public Schools

Beyond enrollment, in 2000, only 74 students received a 3, 4, or 5 on the AP science exams; by 2007, over 290 students received a 3, 4, or 5 (Figure 10.6). Between 2001 and 2007, the number of African American students taking AP science courses (chemistry, physics, and biology) increased by 96% (69 to 135). The number of Hispanic students taking AP science courses (chemistry, physics, and biology) increased by 250% from 18 students in 2001 to 63 students in 2007 (see Figure 10.5).

Figure 10.3

EIGHTH-GRADE SCIENCE STATE ASSESSMENT: PERCENTAGE PASSING

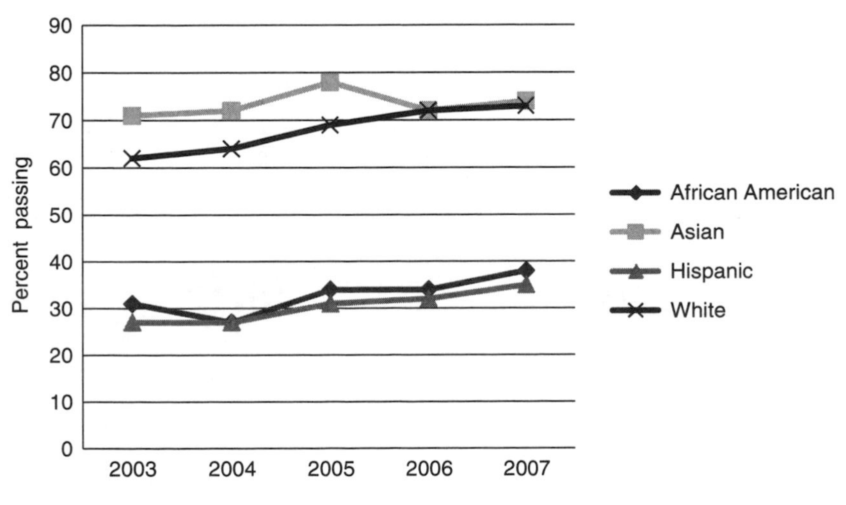

Source: Boston Public Schools

Figure 10.4

AP SCIENCE: GROWTH IN ENROLLMENT

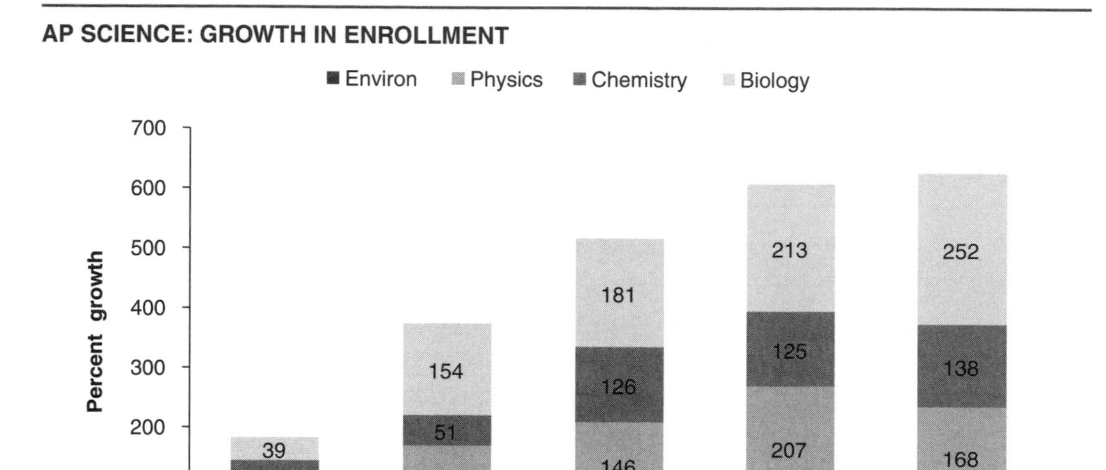

Source: Boston Public Schools

Figure 10.5

AP SCIENCE: PERCENTAGE PASSING WITH A SCORE OF 3, 4, OR 5

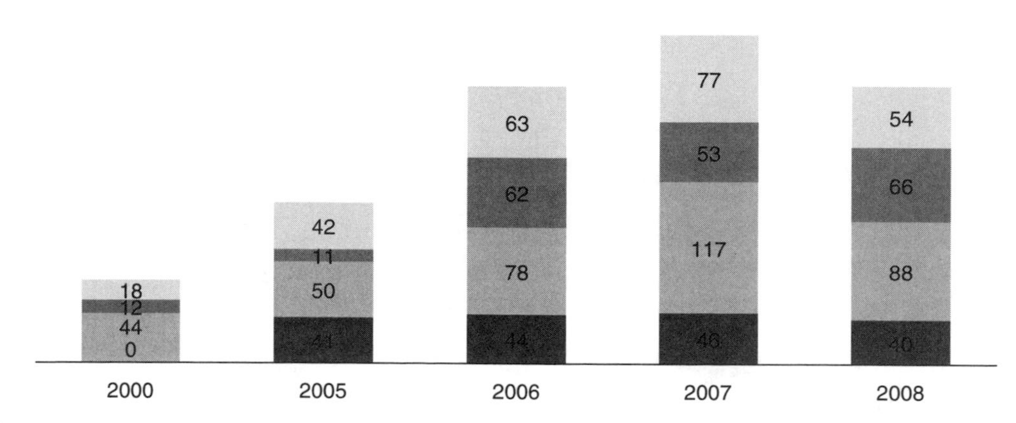

Source: Boston Public Schools

The Trial Urban District Assessment of the National Assessment of Educational Progress showed Boston making greater gains than any other urban district included in the analysis. From 2003 to 2007, the average scale scores in Boston Public Schools increased by 13 points at grade 4 and 14 points at grade 8—both more than double the increase for the average for all large urban cities (6 and 7 points, respectively). From 2003 to 2007, the average scale score in grade 4 increased for all racial groups. The average scale score increased across all racial groups, except Asian students, for the same time period in grade 8. From 2003 to 2007, Hispanic students' gains were among the highest in grade 4 (+15 points) and outpaced all other racial groups in grade 8, with an increase of 18 points.

Lessons Learned

While the results of Focus on Science were impressive, implementation gaps existed within and across schools that caused the initiative to fall short of its full potential goals. Two key factors played a role: the lack of a binding, statewide assessment in science (which did not take full effect until 2010) and the lack of a relentless, systemwide application of the assessment and curriculum implementation review results in every school in the district.

The greatest contributions of Focus on Science were that it helped the districts to do the following:

- implement an inquiry-based science curriculum in grades K–12
- develop strong science leadership capacity within the district
- create and support a districtwide science resource center
- develop a culture of professional development for all teachers of science
- implement an assessment system in grades 6–12
- increase advanced science offerings for students
- develop stronger partnerships with universities and educational entities
- increase (drastically) the number of certified physics teachers
- increase teacher retention in science

Reflections for the Future

Even though new policies, monitoring practices, support services, and an adequate budget were in place, the importance of generating and sustaining a world-class science program, with world-class student results, was not as high a priority for the district as we had hoped. Binding, high-stakes state assessments in English language arts and mathematics that placed the district and schools in jeopardy of substantial sanctions and public humiliation if students performed poorly grabbed the attention of school and district leaders and held them hostage.

The districtwide science assessment and Curriculum Implementation Reviews results were shared with top-level district officials in semiannual and individual meetings and generated substantial interest and concern; however, that interest and concern never translated into the relentless monitoring of school practices in the areas of curriculum and instruction that would have improved the science performance of students, teachers, and schools. If change were going to take place and be sustained, it would have to happen from the bottom up. Although this kind of change did come about in classrooms and schools throughout the district, it did not come to all schools nor, unfortunately, to all students during the time of the reform efforts described in this chapter. Thus, as we reflect on the future, we know our work continues.

References

Davis, K. 2002. "Change is hard": What science teachers are telling us about reform and teacher learning of innovative practices. *Science Education* 87 (1): 3–30.

Duschl, R. A., H. A. Schweingruber, and A. W. Shouse, eds. 2007. *Taking science to school: Learning and teaching science in grades K–8.* Washington, DC: National Academies Press.

Fiarman, S. E., S. M. Johnson, M. S. Munger, J. P. Papay, and E. K. Qazilbash. 2009. Teachers leading teachers: The experiences of peer assistance and review consulting teachers. Paper presented at the meeting of the American Educational Research Association. San Diego, CA.

Fields, E. T., A. J. Levy, M. K. Tzur, A. Martinez-Gudapakkam, and E. Jablonski. 2012. The science of professional development. *Phi Delta Kappan* 93 (8): 44–46.

Guskey, T. R., and K. S. Yoon. 2009. What works in professional development? *Phi Delta Kappan* 90 (7): 495–500.

Lord, B., K. Cress, and B. Miller. 2008. Teacher leadership in support of large-scale mathematics and science education reform. In *Effective teacher leadership: Using research to inform and reform,* ed. M. M. Mangin and S. R. Stoelinga. New York: Teachers College Press.

Loucks-Horsley, S., N. Love, K. E. Stiles, S. Mundry, and P. W. Hewson. 2003. *Designing professional development for teachers of science and mathematics.* 2nd ed. Thousand Oaks, CA: Corwin Press.

Metz, K. 2008. Elementary school teachers as "targets and agents of change:" Teachers' learning in interaction with reform science curriculum. *Science Teacher Education* 93 (5): 915–954.

Spillane, J. P. 2000. District leaders' perception of teacher learning. CPRE occasional paper series, OP-05. Philadelphia, PA: Consortium for Policy Research in Education.

Spillane, J. P., and K. A. Callahan. 2000. Implementing state standards for science education: What district policymakers make of the hoopla. *Journal of Research in Science Teaching* 37 (5): 401–425.

President's Council of Advisors on Science and Technology (PCAST). 2012. Report to the President. Engage to excel: Producing one million additional college graduates with degrees in science, technology, engineering, and mathematics. *http://www.whitehouse.gov/sites/default/files/microsites/ostp/pcast-engage-to-excel-final_2-25-12.pdf.*

U. S. Department of Education (USED). 2010. A blueprint for reform: Reauthorization of Elementary and Secondary Education Act. U.S. Department of Education. *www2.ed.gov/policy/elsec/leg/blueprint/blueprint.pdf.*

Yin, R. 2006. *Cross-site evaluation of the Urban Systemic Program. The final annual report: Baseline outcome analysis.* Washington, DC: COSMOS Corporation.

Chapter 11

Seattle Public Schools' Professional Development Model: Preparing Elementary Teachers for Science Instruction

Elaine Woo

Rationale

Seattle Public Schools' preK–12 science staff has been collaborating with district and community leaders since 1994 to bring about systemic change in how science teachers are supported and how science is taught so that students have the opportunity to achieve at high levels. Over the years, these science education stakeholders developed a mission statement: "All students are able to investigate scientifically in order to construct and acquire conceptual understanding of their world, develop positive scientific attitudes, and become scientifically literate." This is accomplished through a collaborative, interactive, rigorous science program responsive to the needs of diverse learners. Through dedication, persistence, and attention to research and detail, Seattle's science staff has produced a professional development model that provides a system and infrastructure to enact this mission (see Figure 11.1, p. 166)

For more than 16 years, students and teachers in our state have benefited from grant funds from the National Science Foundation (NSF), the state-level Mathematics and Science Partnership (MSP), and Washington State LASER (Leadership and Assistance for Science Education Reform) for pursuing science education reform. Seattle Public Schools received a Local Systemic Change (LSC) grant from NSF in August 1996 for the purpose of bringing systemic change to elementary science education. In 1998, the district became involved in a regional LSC grant for middle schools with five districts, followed by a second research grant. Eleven years ago, preK teachers were included in the reform efforts because of teacher demand. Finally, about seven years ago, the district started receiving funding to support professional development and systemic change at the high school level.

The Puget Sound area has many scientific research institutions and industries whose leaders and employees care deeply about student achievement in preK–12 science. The support provided by individuals and outreach groups from these institutions has complemented the support as detailed above and has allowed concentrated reform work over many years (see Figure 11.2, p. 167).

Figure 11.1

DISTRICT-COMMUNITY PROFESSIONAL DEVELOPMENT SUPPORT INFRASTRUCTURE LOGIC MODEL

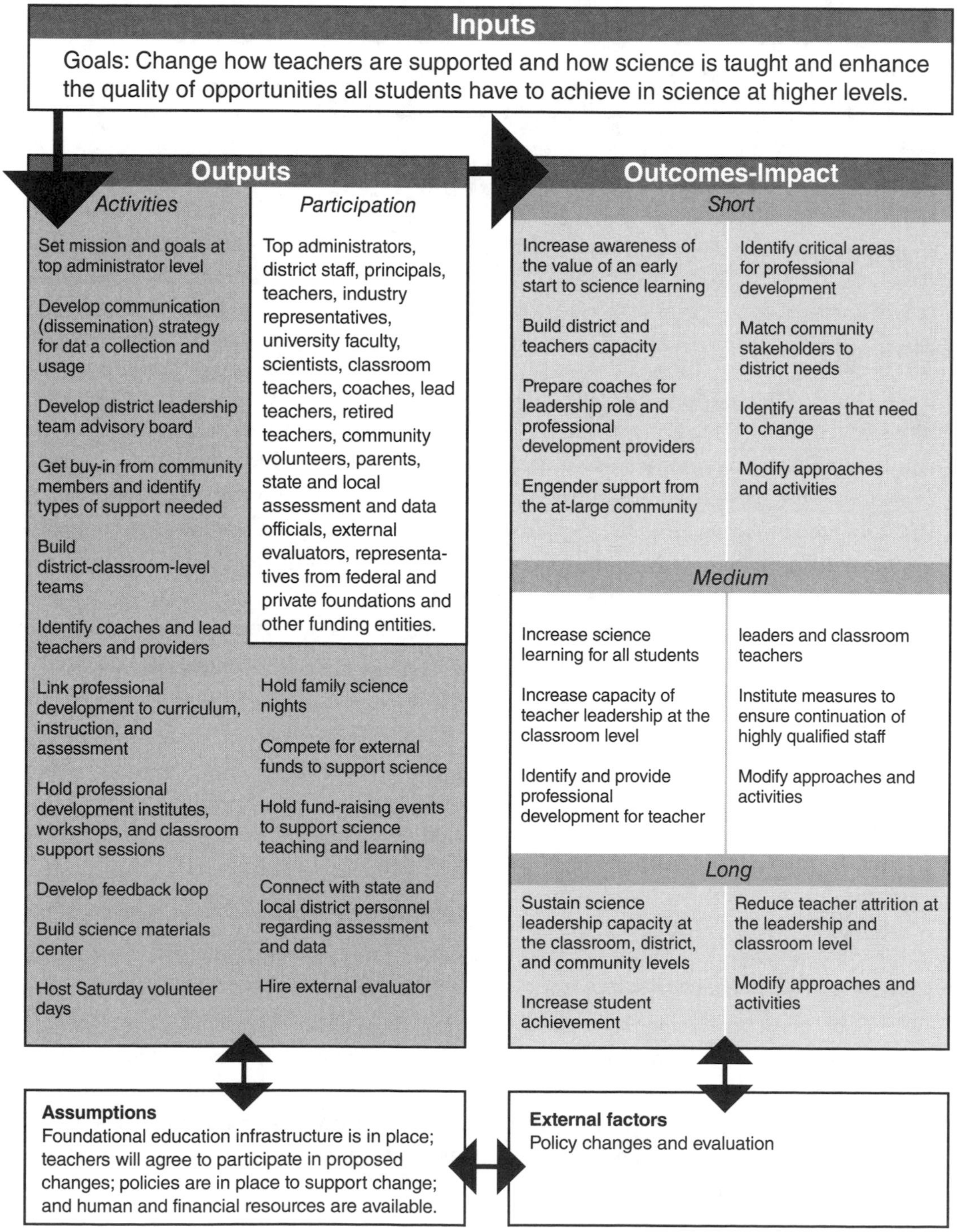

Inputs

Goals: Change how teachers are supported and how science is taught and enhance the quality of opportunities all students have to achieve in science at higher levels.

Outputs

Activities

Set mission and goals at top administrator level

Develop communication (dissemination) strategy for dat a collection and usage

Develop district leadership team advisory board

Get buy-in from community members and identify types of support needed

Build district-classroom-level teams

Identify coaches and lead teachers and providers

Link professional development to curriculum, instruction, and assessment

Hold professional development institutes, workshops, and classroom support sessions

Develop feedback loop

Build science materials center

Host Saturday volunteer days

Hold family science nights

Compete for external funds to support science

Hold fund-raising events to support science teaching and learning

Connect with state and local district personnel regarding assessment and data

Hire external evaluator

Participation

Top administrators, district staff, principals, teachers, industry representatives, university faculty, scientists, classroom teachers, coaches, lead teachers, retired teachers, community volunteers, parents, state and local assessment and data officials, external evaluators, representatives from federal and private foundations and other funding entities.

Outcomes-Impact

Short

Increase awareness of the value of an early start to science learning

Build district and teachers capacity

Prepare coaches for leadership role and professional development providers

Engender support from the at-large community

Identify critical areas for professional development

Match community stakeholders to district needs

Identify areas that need to change

Modify approaches and activities

Medium

Increase science learning for all students

Increase capacity of teacher leadership at the classroom level

Identify and provide professional development for teacher leaders and classroom teachers

Institute measures to ensure continuation of highly qualified staff

Modify approaches and activities

Long

Sustain science leadership capacity at the classroom, district, and community levels

Increase student achievement

Reduce teacher attrition at the leadership and classroom level

Modify approaches and activities

Assumptions
Foundational education infrastructure is in place; teachers will agree to participate in proposed changes; policies are in place to support change; and human and financial resources are available.

External factors
Policy changes and evaluation

Figure 11.2

PROFESSIONAL DEVELOPMENT: TEACHER-CLASSROOM IMPACT LOOP

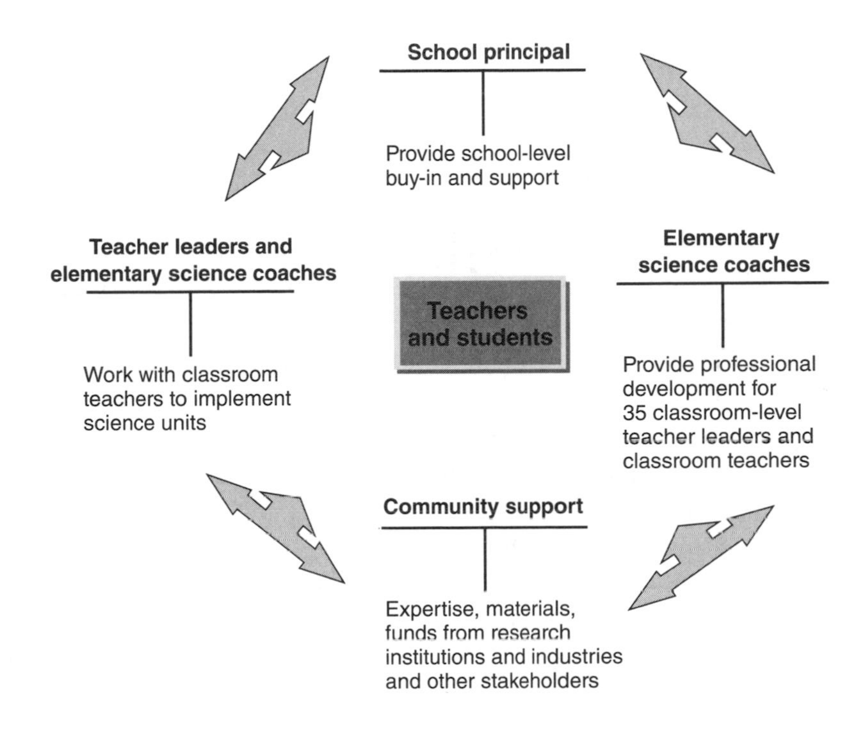

As a part of our support infrastructure, centralized science coaches at each of the four levels (preK, elementary, middle school, and high school) focus their efforts in slightly different ways based on the needs of the students and teachers. The science coaches and program manager have had a disciplined focus on evolving the quality of professional development and curricular support for teachers so that students have a foundation for becoming scientifically literate, gain the skills and knowledge necessary to be successful on the state science assessment, and become college and workforce ready. This chapter concentrates on the system and infrastructure that was developed for supporting teachers and students at the elementary level. The elementary science educators have been involved in reform efforts the longest, with approximately 1,000 to 1,200 elementary teachers in the district supported each year by the program, thereby allowing extensive practice over time for developing specific program components.

Laying the Foundation for the Model

The district science coaches and program manager, collaborating with their evaluators and community scientists, were able to develop an effective model of professional development because in the early years of the program important key elements were put into place, and over time these key elements evolved into lessons learned that we can pass to others. The staff has continued to develop and refine these elements, and they have helped us move forward and evolve our professional development system to provide strong support for students and teachers. The key elements included the following:

A Focused, Common Vision Among Stakeholders

The first leadership team or advisory board was made up of the program manager; one science coach; one to two principals; one supervisor of principals; a university science educator; and representatives from industry, the citywide Parent-Teacher-Student Association, and the Alliance for Education, a community organization supporting the mission of the school district. This board met about six times a year to support the development of the vision, brainstorm ideas for growth, and troubleshoot challenges. The original board dissolved after the first five years, and a different group made up of similar stakeholders was convened about three years later. These boards come and go depending on need. It is very important to have these groups reconstitute and reconvene as the program moves into new phases. The refinement of the vision and infusion of new ideas is vital for the growth and sustainability of a complex, multifaceted systemic program.

The original district science reform vision was developed at the summer LASER conference at the National Science Resources Center (recently renamed the Smithsonian Science Education Center) in Washington, D.C., in 1995 and has been refined and distilled over the years with input from various stakeholders to "All students become scientifically literate."

Strong Collaboration Among and Support From Community Partners, Including Scientists

The collaboration with community partners and scientists goes well beyond the leadership team. Community partners can help with grant writing and seeking resources and funding. For example, university partners and staff from science research institutions, such as the Institute for Systems Biology, the University of Washington, Seattle Pacific University, and the Alliance for Education, have helped us write proposals.

Partners can also help with the logistics and technicalities of preparing science kits and with other materials supports issues. For example, Seattle Works is a group that provides a structure in which people in their 20s and 30s find volunteer opportunities in their communities. We have had many Saturday volunteers from this group help prepare materials and science kits for classroom use. This has become a way for our program to share with the public the positive and productive work going on in our district to support student learning. Our Science Materials Center also benefits from volunteers from the annual United Way's Day of Caring. As another example, a retired high school physics teacher and his wife have worked with our science coaches for 18 years to troubleshoot materials, equipment, and packaging challenges, all of which impact professional development and instruction in the classroom. Community partners can help organize Family Science Nights and other special events to enhance science learning and help parents understand what students are learning in science. The Institute for Systems Biology provided substantial support in helping schools develop strategies and procedures to organize Family Science Nights, and institutions such as the Woodland Park Zoo offered special attractions at these parent events by bringing animals and displays. Finally, it is crucial to work with university science educators on content issues for all aspects of the program and professional development.

Continuity of Highly Qualified Staff

It is typical to have coaches or program managers change positions every two to five years, but it is difficult to build strong programs if there is too much turnover. Our goal is to develop a continuity of staff to maintain a strong, common vision and to have continuous growth of the program and professional development. Having community partners with strong convictions to provide support politically and in fund-raising helps with the maintenance of highly qualified staff.

Ongoing Professional Development for the Science Consultants and Other Staff

It is important that our staff participate in ongoing professional development. We all need to learn and improve our skills in order to serve the students and teachers better. The entire staff participated in both general district-supported professional development for central office staff and also science education–specific professional development. During the LSC grant program, the elementary science coaches each averaged 80 hours of professional development in each of the five years of the award. They participated in courses on relevant content, inquiry, professional development design, facilitation skills, coaching skills, assessment, and science education reform. Mutual participation in staff development sends the message to the teachers that we are in this together, and we all need professional development in order to improve our skills and do our jobs better.

A Credible Feedback Loop for Teacher Input

Teachers must have the opportunity to give feedback at the end of each class or workshop, and that feedback must be heard. The science staff must take teacher feedback seriously and adjust professional development based on the needs of the participants. At the same time, the coaches must shape the courses on research and best practice. Science coaches learn about the needs of the teachers through teacher feedback and classroom observations and tailor courses to be useful and effective for our teachers. When the coaches attended their own professional development, they synthesized what they had learned and integrated it into existing classes (professional development offerings). When courses can evolve in response to teacher input over time as well as research and national best practices, these courses and workshops continually reach a higher level of quality of support for teachers and, in turn, allow teachers to improve their instruction and support higher levels of student learning and interest in science. The final LSC evaluation report from Inverness (St. John 2002) made note of the high quality of professional development experiences that were provided for teachers in support of improved science instruction across the district.

Use of Current Data, Research, and Best Practices

It is essential that the science staff use data from classroom observations, formative assessments, science notebooks, state assessments, research, and best practices. All this information must be considered as the staff develops, shapes, and refines professional development. Again, the courses must meet the needs of teachers and provide them with approaches and strategies that are proven through research and field testing.

Discipline

In his *Good to Great and the Social Sectors: A Monograph to Accompany Good to Great*, Jim Collins talks about "the relentless culture of discipline" (Collins 2005, p. 1). There is mention of "disciplined planning, disciplined people, disciplined governance, disciplined allocation of resources" (p. 1). As the elementary science program was developed, the science staff took on an unstated culture of discipline. Because we raised the bar high for ourselves, for teachers, and for students, we believe we changed the level of our outcomes. Educators often talk about vision, collaboration, and use of data, but they do not always talk about the use of discipline. Discipline is a concept that emerged in our program unexpectedly, and we now believe it to be critical for providing the best support for students and teachers.

Belief in Taking Time to Fully Develop Cutting-Edge Program Components

With both the science unit instruction and the implementation and integration of the science writing approach, when coaches make a change with materials or in the instructional guides, it is most often piloted in one or more classrooms with strong teachers. The coach and teacher analyze how the change works and make refinements to improve outcomes. Only when they are satisfied are the new materials and changes shared with all teachers and used to improve other courses.

An Effective Outside Evaluator

Having an outside evaluator is usually dependent on having outside funding. When it is possible, outside input and guidance can be extremely valuable. Evaluators work with staff in many districts, and they look at what educators are doing based on a broad context. They can see patterns that work and patterns that do not. When local staff is closely focused on the work of the project, it is sometimes difficult to see ineffective steps that an evaluator can see with clarity. In addition, these researchers provide an informal kind of professional development; we learn from them. Having these advisors is invaluable when first developing a program and over the years for refining the program.

Defining the Professional Development System

The Seattle professional development support system is a legacy of the NSF LSC grant work. We continue to use the inquiry-based science model developed during the LSC and refine the courses and workshops we offer each year. Science coaches developed courses with the overall goal of providing teachers with support so that all students gain the foundation by the end of the fifth grade to be successful in middle school science. Initial Use (a required class presenting inquiry-based science units) and science writing courses are the two most basic courses classroom teachers attend. Teachers are required to take the six- to nine-hour Initial Use classes in order to receive science units for their grade levels. (Three intermediate Initial Use classes (professional development offerings) are nine hours because of the complexity of the units and because of teacher demand to lengthen the class.) State assessment prep and science content classes are optional. The classes we offer each year vary depending on the funding we have for their support.

Initial Use classes

To sustain the Initial Use classes after the LSC grant, the evaluators said that it was essential for the science coaches to develop a team of lead teachers who would help teach Initial Use classes. The limited number of science coaches that might be funded would not be able to put on the number of Initial Use classes needed. The science coaches have been released from the classroom for several years and have had a lot of professional development of their own as mentioned above. At the peak of the LSC grant, we had eight elementary science coaches. The first year after the grant, we had two science coaches. In subsequent years, we have had 2.3–3.3 elementary science coaches to serve the approximately 1,000–1,200 elementary teachers in our district.

At first, it seemed impossible for the science coaches to prepare the approximately 35 lead teachers needed to teach the classes, and it took about three years to get this number of lead teachers in place. The lead teachers are classroom teachers who come after school to teach Initial Use classes. Two, and sometimes three, lead teachers work together to teach the class for each science unit.

Before the lead teachers take over the instruction of a unit, coaches work with them and model several sessions. We found it desirable for the coaches to model for two years (four sessions) before the lead teachers began to instruct independently. We consider this to be a quasi-apprenticeship model. Before the classes, the lead teachers meet with a science coach to go over the plan for the class, learn facilitation strategies for working with adult learners, and review science concepts and practices. They meet again after the session to debrief what went well and what might be changed.

During the Initial Use class, teachers receive an extensive instructional guide along with other supportive documents and later the kit of materials to support the unit and a teacher manual. In the class, teachers experience the lessons from the unit as a student would.

In an Initial Use class, teachers

- become familiar with the scientific content in the unit;
- become familiar with all the lessons and how they build to develop conceptual understanding;
- learn how to use the Learning Cycle with each lesson when planning and implementing inquiry-based science lessons;

- deepen understanding of inquiry by actively participating in key lessons;
- become familiar with the instructional guide, which has been developed specifically for each science unit and provides modifications, extensions, and additional lessons in order to address the state standards and performance expectations;
- learn about classroom-based assessment strategies; and
- become familiar with materials and management strategies to ensure success with the unit.

The district's science coaches have partnered with scientists to develop the instructional guides mentioned in the fifth bullet above. These guides are meant to support teachers in the classroom as they implement each science unit. The instructional guides include

- instructional modifications meant to focus learning on the standards, such as strategies for introducing systems thinking, emphasizing the planning of controlled investigations, and introducing specific vocabulary from the state performance expectation.
- clarifications and extensions of the teacher manual for each unit.
- tips for each lesson in the unit that provide ideas for use of materials, introduction of extended concepts, and use of strategies to promote skill development. Science coaches have developed these tips by observing students doing investigations in classrooms, working with scientists, and incorporating national-level research on best practices.
- additional lessons, developed by consultants and scientists, provide teachers with a lesson plan in which the Learning Cycle is explicit. An example is a lesson supporting teachers that focuses on fair tests or controlled investigations. Planning fair tests begins on a very basic level in kindergarten and grows in complexity in successive grades.

In a district the size of Seattle Public Schools, it is difficult to keep classroom teachers aware of new developments and best practices in science, so these instructional guides were developed, in part, for this purpose and are continually updated and sent to teachers during the trimester they teach a unit title. Through these instructional guides, we are able to keep teachers up-to-date on changes with state standards and requirements.

Science Writing Classes and Supplementary Writing Curriculum

In the science writing classes, the developer, who is one of the district's science coaches, explains how to teach expository writing in the context of teaching science and developing and deepening students' scientific thinking and understanding in the process. Through modeling scientific language and thinking and providing different types of visual, oral, and written scaffolding, the teacher supports students as they learn how to think, talk, and write about science. The modeling and scaffolding are provided as a way to help students make meaning of their investigations and talk about their thinking. Then, the modeling and scaffolding support students as they learn to write about their investigations and their thinking (Fulwiler 2007). The coach also models how science notebooks should be used for formative assessment. These classes and the supplementary writing curriculum are part of the science writing program, which a specialized coach has been developing over the last 15 years. The program includes three components:

1. Professional development available to all elementary teachers across the district. This component consists of a series of four workshops per grade level that introduce the overall approach and apply it to each specific unit following suggestions in the supplementary writing curriculum.
2. A supplementary writing curriculum for each of the 18 elementary science units. Each curriculum includes suggestions for integrating science and expository writing instruction in every lesson. Like the workshops, the curriculum explains how to use word banks, graphic organizers, and scaffolding for teaching expository writing and scientific thinking.
3. Teacher leadership development for 6–10 teachers per grade level. These lead science writing teachers assist in developing and field testing supplementary curriculum strands and materials for other elementary teachers, provide writing samples to be used in professional development and supplementary writing curriculums, and improve their own instructional practices by planning lessons and analyzing student notebooks with their colleagues using protocols developed in the program.

When students begin a science lesson, they take out their notebooks, write the date at the top of the next page, and when appropriate, write the focus or investigation question for the lesson. These questions keep both the students and teachers focused on the conceptual storyline of the unit. After an introductory class discussion, students may write their prediction about the outcomes of the session's investigation. Then, as they conduct their investigation, students collect and record qualitative and/or quantitative data. Students then discuss and analyze their data during reflective class discussions about their investigation.

In a separate instructional period, the teacher helps students learn how to write scientifically about their investigation, then provides scaffolding such as sentence starters or graphic organizers to guide students as they write independently in their science notebooks. Several times during the unit, teachers collect the notebooks to analyze entries and provide constructive feedback that focuses on the students' strengths and addresses weaknesses by asking questions that scientists might have about the entries. Through this process, teachers help students learn scientific writing skills as well as develop scientific thinking and understanding.

Inverness Research Associates of Inverness, California, found the following results after four years of intensive research on the science writing program:

- Independent experts judge that the student work in science notebooks is, on the whole, more sophisticated in quality and reflective of greater rigor and a higher level of learning of both science and writing than is typical in science programs in other schools and districts that use similar science units.
- The writing program thus enhances to a significant degree the district's elementary science program, and it helps bolster the district's literacy program, including the extent to which those programs help students meet state standards.
- Participants in the writing program spend more time teaching science, teach more writing in science, have higher expectations for students with special needs, and follow the district's science curriculum more consistently than teachers who have little or no experience with the Expository Writing Program. (Stokes, Hirabayashi, and Ramage 2003, p. 7)

The National Center for Research on Evaluation, Standards, and Student Testing (CRESST) at the University of California in Los Angeles has completed additional research on the impact of the science writing program on students' science learning. They completed a detailed analysis of Seattle's fifth-grade students' state science assessment scores to determine if there is a significant impact when students' teachers have seven and a half hours or more of professional development in the science writing classes. The researchers found that students taught by teachers who had had three days of professional development for an inquiry science unit at the grade level taught (three Initial Use classes) plus over seven and a half hours of classes in the science writing out-performed other students with significantly higher scores (Choi and Herman 2007). Launching a National Professional Development Institute for Science Writing in 2011 allowed Seattle's coaching leadership to serve a national audience (Ramage 2012).

Thus, 10 years of qualitative and quantitative evidence show that the strategies described in the science writing classes and supplementary writing curriculum make a dramatic impact on elementary students' achievement in both science and expository writing. These documented successes of the program are due in part to its substantial and purposeful integration with Initial Use courses focused on the science units. Betsy Rupp Fulwiler, the developer and implementer of the science writing approach, has written a book, *Writing in Science: How to Scaffold Instruction to Support Learning* (2007), describing in detail how teachers can implement this approach. Her second book, *Writing in Science in Action: Strategies, Tools, and Classroom Video* (2011), provides additional support for implementing the science writing approach.

The State Science Standards and State Assessment Prep Classes

Responding to the reauthorization of the Elementary and Secondary Education Act of 2001 (No Child Left Behind), Washington State's Office of the Superintendent of Public Instruction published *Science K–10 Grade Level Expectations: A New Level of Specificity* in 2005. Then, in 2009, they came out with a new and more explicit and rigorous set of standards,

Washington State K–12 Science Learning Standards. There are four science standards: systems, inquiry, application, and the domains of science. Washington is unique in emphasizing "systems" because of its growing importance in diverse fields, such as climate change and genetic engineering (*www.k12.wa.us/Science/pubdocs/WAScienceStandards.pdf* , p. 3). In addition, the big idea of energy became more important in the new state standards, which was a shift from the past and brought a new challenge for teachers.

The state assessment is made up of three types of scenarios: systems scenarios, inquiry scenarios, and design scenarios. A scenario is a short, hypothetical example of what students might encounter in a school investigation or in the world outside of school. The scenarios are based on the language of the standards and performance expectations. The science assessment is made up of at least 20% Systems points, 30% Inquiry points, 20% Application points, and no more than 30% for the Domains of Science. The 30% focus on inquiry is driving the implementation of inquiry throughout the K–12 classrooms in the state, including Seattle's classrooms. The test items include multiple choice, completion items, and short-answer items.

In response, Seattle's State Assessment Class for professional development provides support on systems subjects, during which science coaches model adjusted lessons that incorporate learning about systems. Also, these classes involve addressing the second standard on inquiry, guiding students in planning a fair test or controlled investigation. During the class, the science coach explicitly models for the teachers how to guide students in coming up with testable questions and then plan and conduct fair tests or controlled investigations with identified variables. Finally, the science writing consultant presents instructional writing strategies that support success on the state science assessment. The coach explicitly models how to guide students in interpreting generated data and in writing a conclusion that includes the components required by the state assessment. This coach also provides the teachers with a suggested template that supports this modeling process. The template includes components that are not assessed on the state science assessment but that promote higher-level thinking and more complex expository writing skills.

In the State Assessment Prep classes, the coaches emphasize and show how each grade level contributes to preparing the students for the fifth-grade state assessment. Teachers have reported that the components of this class helped them immensely because the content of the class is tied specifically to the concepts and skills of the science units they are teaching rather than being generalized information.

Earlier, Seattle Public Schools offered separate classes for third-, fourth-, and fifth-grade teachers focusing on strategies embedded in each science unit. We also offered a primary state assessment prep class for K–2 teachers. With a tighter budget, we now offer a State Assessment Prep class for both fourth- and fifth-grade teachers during the day, with half-day substitute release as well as after school for three hours. The feedback was positive with this change.

> I went to the MSP science prep training for fourth- and fifth-grade teachers yesterday. It was very informative re the new 2009 standards that this year's

> science MSP will assess. One of the biggest changes is in the jargon, which we need to change in our teaching so the kids won't be blindsided when they take the test. There are also several new standards, including space science, which they may be tested on. Our curriculum doesn't address space science at all, so [district science coaches] have written a new mini unit for fourth and fifth grade, which is not required but is recommended. The instructional guides, which come with our kits this year, have significant changes to address the new standards. ... There is another MSP science prep class in February, which I would recommend. (Sandra Hannes in an e-mail to her school colleagues, January 2011)

In these classes, teachers receive documents showing adjustments to embed into the instruction of their grade-specific science units. There is a focus on explicitly showing them how to teach and model for student investigations, including the standard of systems and the big idea of energy.

Science Content Courses

For seven years, the elementary science program offered summer content courses taught by scientists with assistance from science coaches and lead teachers for each grade level for the purpose of helping teacher participants understand the content at a higher level than what the students must learn. Science coaches and teacher leaders work with the scientists to make sure that the course aligns with our standards, science units, and instructional approach and to make certain that scientists are aware of the realities of the classroom. These courses are usually from 12 to 24 hours in length.

Over the years, science partners have collaborated with us using grant funds and have offered content courses for a particular grade level. In the summers of 2009–2013, for example, Seattle Pacific University offered our intermediate teachers, mainly in physical science, a content course addressing energy. Our staff value these courses and promote them with our teachers. Meanwhile, the coaches infuse as much content as possible into the Initial Use, the Expository Science Writing, and the state science assessment prep classes. We continually work with scientists as we refine instructional strategies and develop supplementary curriculum for each science unit.

Lessons Learned

How have Seattle's fifth graders done on the state science assessment after many of the teachers experienced the well-developed professional development system as described above? Washington's fifth-grade students started piloting the state science assessment in 2001. The fifth-grade exam became operational in 2004 and mandatory in 2005. The bar graph in Figure 11.3 shows that Seattle's students had a higher average percent proficient in the first six years of the state assessment than other fifth graders in the state. This result is unexpected given the diversity of the student population within Seattle Public Schools. The district is made up of about 49,000 K–12 students, with about 40% receiving free and reduced-price lunch and approximately 61% being students of color. These students speak over 100 different languages. In addition, the program's researchers (CRESST) found the difference between Seattle's fifth graders' scores and the scores of fifth graders statewide to be considerably larger than would be expected: 6.3% more Seattle students met the standard than did students in districts of similar size and demographics (Soholt 2005).

In 2010, the state assessment was changed enough from the first six years for researchers to conclude that the previous years' scores should not be compared with the scores from 2010 forward. What we can say is that although the 2010 elementary scores went down in almost every district in the state, Seattle's fifth graders' average percent proficient was still almost 7% above the state average. This test was shorter, and therefore, each missed item counted for more on the score (Figure 11.4, p. 178) . From 2010 through 2013, Seattle's fifth-grade science assessment scores have continued to rise and remain higher than the fifth-grade scores for the state.

Figure 11.3

IMPACT OF REFORM ON FIFTH-GRADE SCIENCE SCORES ON THE WASHINGTON ASSESSMENT OF STUDENT LEARNING

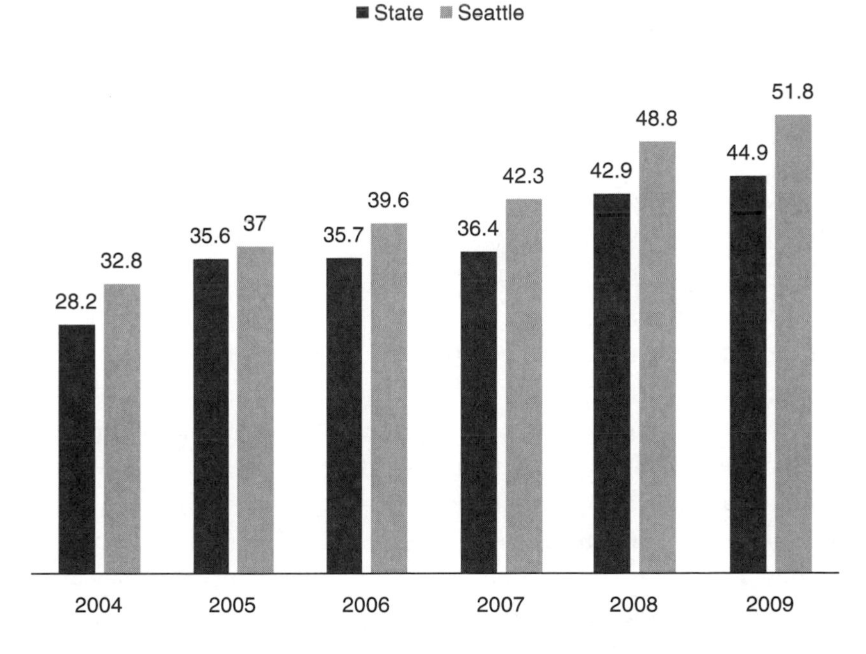

Source: Washington State Office of Superintendent of Public Instruction and Washington State Board of Education

Although achievement is not where we would like it to be, the scores have been improving. Students' science notebooks show strong scientific thinking and conceptual understanding when the science units and science notebooks are both implemented as our science coaches suggest, and the evaluators and researchers say this is not an accident. The ongoing professional development and teachers' efforts have made a positive impact. The

Figure 11.4

IMPACT OF REFORM ON SEATTLE'S SIXTH-GRADE STATE ASSESSMENT SCORES

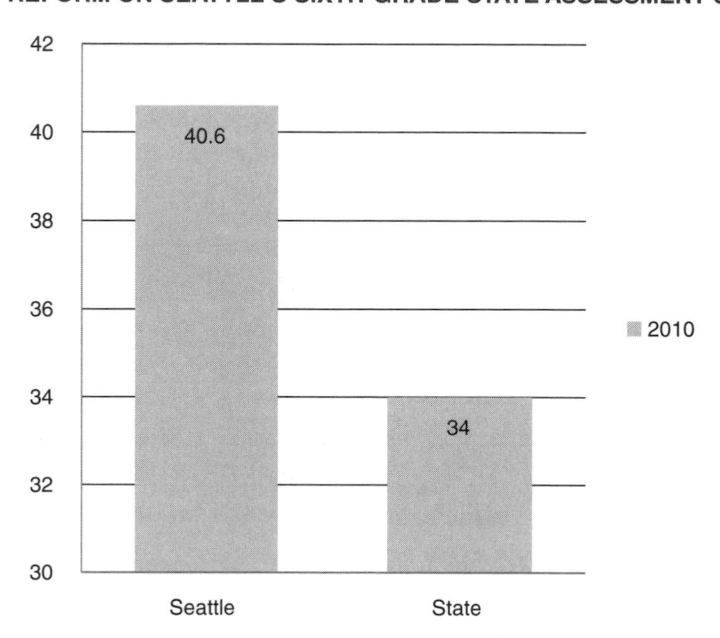

Source: Washington State Office of Superintendent of Public Instruction and Washington State Board of Education

challenge continues to be getting all teachers to teach the science units and implement the science writing approach developed in Seattle.

Reflections for the Future

The elementary science coaches in Seattle Public Schools have been adamant about supporting teachers explicitly in learning how to address specific state standards and performance expectations in their instruction of each science unit. In addition, they support teachers in teaching beyond and at a higher level than what the state assesses. The explicit modeling for the teachers both in professional development classes and workshops and through documents is in contrast to giving teachers general ideas that they then have to apply on their own to their instruction. Elementary teachers need specific, explicit support as they teach multiple subject areas. Evidence of the success of this approach is shown over 10 years of reform efforts with the steady improvement in state science assessment scores and improvement in individual student science notebooks. The professional development reform infrastructure, which was developed over many years, has clearly allowed Seattle Public Schools' science staff to develop and refine an effective professional development program model in inquiry-based science. The support provided by this program helps teachers to ensure that students reach learning goals and are well prepared for middle school science.

Acknowledgments

Major partners in Seattle's professional development model implementation included the NSF and the Charles Simonyi Fund for Arts and Sciences. In addition, the work has been supported by Seattle's Alliance for Education, Amgen, The Boeing Company, the City of Seattle: the Seattle Aquarium and Seattle Public Utilities, Commonweal Foundation, George Rathmann Foundation, Institute for Systems Biology, Medtronic, Mental Wellness Foundation, Nesholm Family Foundation, Washington State LASER, and ZymoGenetics. An early phase of the science writing work was funded by Social Venture Partners. Seattle Pacific University, Department of Physics; University of Washington, Department of Biostatistics, Department of Medicine, Division of Medical Genetics, and the Physics Department also support our professional development efforts.

Any opinions, findings, and conclusions or recommendations expressed in this chapter are those of the author and do not necessarily reflect those of the foundations or institutions who were our partners.

References

Banilower, E.R., P. S. Smith, I. R. Weiss, K. A. Malzahn, K. M. Campbell, and A.M. Weis. 2013. Report of the 2012 National Survey of Science and Mathematics Education. Chapel Hill, N.C.: Horizon Research.

Choi, K., and J. Herman. 2007. Seattle school district Expository Writing and Science Notebooks Program: Using existing data to explore program effects on students' science learning. Los Angeles, CA: UCLA National Center for Research on Evaluation, Standards, and Student Testing (CRESST).

Collins, J. 2005. *Good to great and the social sectors: A monograph to accompany good to great.* Boulder, CO: Jim Collins.

Fulwiler, B. R. 2007. *Writing in science: How to scaffold instruction to support learning.* Portsmouth, NH: Heinemann.

Fulwiler, B. R. 2011. *Writing in science in action: Strategies, tools, and classroom video.* Portsmouth, NH: Heinemann.

Office of Superintendent of Public Instruction. 2010. *Washington state K–12 learning standards.* Olympia, WA: Washington State Office of the Superintendent of Public Instruction.

Ramage, K., and L. Stokes, with assistance from H. Mitchell. 2012. Helping students learn science through writing and writing through science: Key findings from ten years of study. Inverness, CA: Inverness Research.

Show, K., and E. Woo. 2008. Washington state's science assessment system. In *Assessing science learning: Perspectives from research and practice*, ed. J. Coffee, R. Douglas, and C. Stearns, 357–377. Arlington, VA: NSTA Press.

Soholt, S. 2005. Seattle science expository writing and notebooks [unpublished technical report]. San Francisco, CA: KSA Plus Communications for the Stuart Foundation.

Stokes, L., J. Hirabayashi, and K. Ramage. 2003. *Writing for science and science for writing: A study of the Seattle elementary Science Expository Writing and Science Notebooks Program as a model for classrooms and districts.* Iverness, CA: Inverness Research Associates.

St. John, M., K.A. Fuller, and P. Tambe. 2002. *Seattle Partnership for Inquiry-Based Science: A local systemic change initiative: End of project report*. Inverness, CA: Inverness Research.

Chapter 12

New York City STEM Professional Development Partnership Model

Nancy (Anne) Degnan

Rationale

This chapter describes a progression of strategies, approaches, challenges, and changes in a professional development model for providing secondary science, technology, engineering, and mathematics (STEM) teachers with the knowledge and skills needed to raise achievement for students enrolled in low socioeconomic schools in New York City. The backbone of this professional development model is the Center for Environmental Research and Conservation (CERC) of the Earth Institute at Columbia University. CERC engages in research, applied research, and education programs in pursuit of fulfilling its mission of building environmental leadership and solving complex environmental problems. Besides being a center, CERC is also a consortium of leading conservation and cultural institutions (including the New York Botanical Garden, the American Museum of Natural History, and the Wildlife Conservation Society, EcoHealth Alliance) and is an affiliate of the Department of Ecology, Evolution, and Environmental Biology at Columbia. As both a center and a consortium, CERC scientists, researchers, and practitioners work together at the nexus of environmental, economic, and social considerations of environmental sustainability. Beyond this, they also engage with a rich array of partners to provide professional development activities for science teachers.

CERC has designed and delivered professional development for teachers through formal educational settings based on its internal policy perspective that higher and secondary education should partner together to support the education of all students. To do so, the center and its partners both within and outside Columbia engage in the exchange of big ideas and practices to maximize student achievement as well as their socioeconomic and environmental well-being.

CERC's internal policy perspective dovetails with that of the NYC Department of Education policy regarding the ChildrenFirst reform that grew out of the No Child Left Behind legislation. ChildrenFirst seeks to both improve achievement across all schools and to address persistently low performing schools by moving innovation and effective school change throughout the system. CERC's overarching goal for professional development is to promulgate sound STEM education and STEM information and communication technologies (ICT) teaching and learning with long-lasting benefits for teachers and their students. This goal is shared by CERC's affiliate partners within and outside Columbia and is articulated as the foundation for extending the partnership with NYC public schools. The fundamental premise is that creating engaging and active professional development for teachers has a multiplier impact for NYC's public school students. More, this approach is beneficial, prima facie, for other communities, states, regions, and the nation.

Model

Original Model

The original model for professional development was teacher centered (see Figure 12.1). The model was based on the National Science Education Standards (NSES) for professional development articulated by the National Research Council (NRC 1996). Through this model, CERC's goal was to move teachers beyond what they accomplish in their regular professional development coursework to the transfer of knowledge and skills to actual classroom settings. While CERC was interested in providing professional development that transformed teaching and learning on a continuous basis, it stopped short of input on pedagogy.

Figure 12.1

EARLY TEACHER-CENTERED MODEL OF PROFESSIONAL DEVELOPMENT

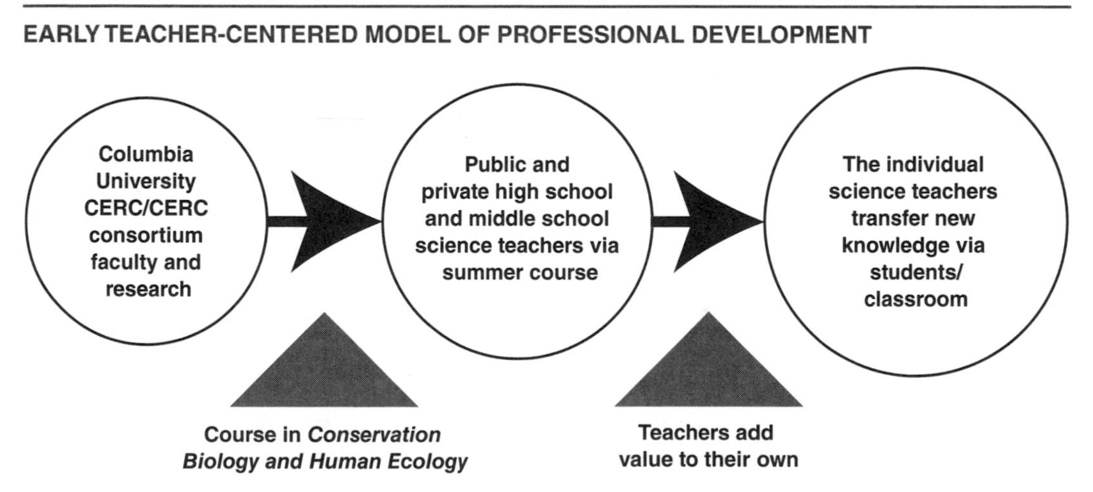

With a direct approach of working with teachers, CERC trained hundreds of high school teachers who impacted thousands of their students during the late 1990s and early years of the 21st century. During those years, professional development involved summer programming for New York State certification in science and specialized workshops focused on topics in conservation science. Teachers also received up to 6 points of graduate credit in conservation biology or human ecology. These points could be applied to the 30 points and above required at the master's degree level or for salary increases in the NYC public school system.

The programming was somewhat distinct from environmental studies inasmuch as it was based in the science of ecology and used conservation as a lens into the environment. Emphasis was placed on exposing teachers to as much information as possible. CERC's viewpoint at that time was that teachers knew how to teach science at the secondary level and that pedagogy from higher education would most likely be a distraction. What was needed, as far as CERC could ascertain, was a richness of content to enhance existing teaching practices and resources.

As a consequence, CERC engaged in little discussion about how to effectively use turnkey information within schools once teachers returned to their classrooms. Although it is highly likely that these discussions occurred among the teachers in schools in an informal way, the transfer of pedagogical knowledge through CERC professional development was not part of the design. In terms of the profile of the participants in CERC professional development, the engagement with teachers was based on traditional recruitment mechanisms of individual applicants to a summer program. Most teachers were exclusively public high school science teachers, but there were participants from independent or private schools as well.

Once the programming was complete, CERC faculty, staff, and teachers had little to no interaction together during the school year except at special events or gatherings.

The Evolution

An evolution in CERC's professional development occurred about midway through the first decade of the 21st century, when CERC entered into a new stage of STEM and STEM information and communications technologies (ICT) engagement with middle school teachers, first through a partnership with a large NYC-based nonprofit and then with the National Science Foundation Innovative Technology Experiences for Students and Teachers division. Additionally, this evolution was marked by CERC's commitment to environmental sustainability. Here, the mission and philosophy were deeply embedded in the belief that it is only through STEM and STEM ICT that CERC would be able to influence the socioeconomic and environmental well-being of current and future generations. Thus, issues of preserving and conserving marine and freshwater resources; transitioning from nearly total dependency on fossil fuel to renewable energy; stewarding of the natural infrastructure—the biodiversity that is life itself—for food, medicines, and other ecosystem services while maintaining a quality of life for ourselves and extending it to others would take all of CERC's ingenuity, know-how, and commitment. STEM became the cornerstone on which CERC's aspirations and those of the education community rest. Specifically, for CERC, STEM was realized through ecology, which was itself integrative of a broad range of knowledge and skills, including chemistry, Earth science, statistics, mathematics, and biology.

Through this professional development, CERC would also seek to honor each school as its own, unique learning community in which CERC strived to be organizationally vigilant about what it knew, what it thought it knew, and what it did not know as a professional development provider. This approach would require CERC to spend significant time in the schools with the principals and faculty to really get to know and understand the local educational process. It would also require CERC to think deeply about continuous improvements to its professional development model to assist in achieving high quality and excellence for all students in the immediate and longer term. The objective was to remain value-added for colleagues in STEM and STEM ICT at the high school level.

This evolution is characterized by three primary transition points (see Figure 12.2) that allowed CERC to rethink its approach to becoming an effective professional development provider given the lack of reflection, follow-up, and support in the original model. The transition points are briefly described in Figure 12.2.

Figure 12.2

THE PROGRESSION OF CHANGES IN CERC PROFESSIONAL DEVELOPMENT

Transition Point 1	Transition Point 2	Transition Point 3
CERC's exclusive focus on Title I middle schools resulting from a New York City–based foundation interested in innovative inquiry-driven and project-based learning education programs as tools to help alleviate poverty in urban schools.	CERC's use or results from lessons learned through work with a large Title I middle school school on an innovative inquiry-driven and project-based learning program produces I-STEM and IPW.	CERC's TREES blended model of lessons-learned from Title I schools, SEE-U, IPW, and I-STEM in conjunction with a two-week professional development course that relies heavily on principal leadership and integrates with ongoing scientific research.

IPW = Integrated Project Week; I-STEM = integrated STEM curriculum; SEE-U = Summer Ecosystem Experience for Undergraduates; TREES = Technology, Research, Ecology Exchange for Students.

Through the partnership with and encouragement of a large nonprofit organization that seeks to positively impact socioeconomic status through public education, CERC changed its focus from directly working with teachers through summer workshops to a middle school initiative that focuses on three areas: basic scientific principles, key STEM concepts in the environmental sciences, and project-based learning. The revised approach to professional development was based on the adaptation of an intensive fieldwork course that CERC operated for nonscience majors at Columbia University, information and perspectives in middle school core curriculum design from the National Middle Schools Association (NMSA), and a review of lessons learned from early years of providing summer workshops for teachers—most notably that principals are a critical partner in professional development.

Adaptation of Undergraduate Field Ecology Course for Professional Development

First, the course for nonmajors, entitled Summer Ecosystem Experience for Undergraduates (SEE-U), had excellent evaluations from students who reported that the program had transformed their perceptions about and sense of self-efficacy in science, specifically field ecology. A small percentage of students actually shifted their majors from social science to natural science. Using data from student evaluations and other measures, CERC concluded that the dual approach of visual and kinesthetic teaching and learning, emblematic of ecological field methods courses, was an important contributing factor to the overall success of the undergraduate program. Hence, with encouragement from a large nonprofit that was interested in finding and supporting programs capable of reversing upward trends of high school dropout rates, the transition was made to focus on the middle school level to test how well an adapted SEE-U program would work for middle school students.

A Focus on Middle School Integrated Core Curriculum

The second aspect of the revised approach came from the NMSA perspective on how to enhance the core curriculum of the middle level grades. In 2002 NMSA endorsed integrated curriculum (basically interdisciplinary curriculum) as foundational for student-centered middle school learning and teaching.

The large nonprofit had asked CERC to focus on the middle school level because research shows that students make decisions during their middle school years as to whether or not to remain in school to complete their high school education (Kurlaender, Reardon, and Jackson 2008). In addition, it is at the middle school level that failing a single course might substantially increase the likelihood of dropping out of high school (Meece and Eccles 2008). Even more studies show that receiving failing grades on standardized mathematics and science tests at the middle school level is linked to poor student performance in the 11th grade (Kurlaender, Reardon, and Jackson 2008).

Beyond those factors, a student's attendance rate in middle school is an important indicator of success in and completion of high school. Finally, leading lives of socioeconomic wellbeing is correlated to having a high school diploma (Bridgeland, Dilulio, and Morison 2006; Acemoglu and Angrist 2000; Rumberger 1983). All of these factors convinced CERC that working at the middle school level could have powerful implications for student success and performance. At this point, the question for CERC was, "How can we help make a middle school curriculum so compelling that students from economically disadvantaged backgrounds will want to attend school regularly and thus enter and complete high school?"

Lessons Learned

The Principal Is Key

With the knowledge gained through prior efforts and partnerships, CERC recognized that making the change to middle school would not be possible without significant buy-in and support from principals. Principal leadership and support set the tone for construction of decision making and engagement of teachers and students. To ensure that such opportunities exist, and with full knowledge that principals have little free time during working hours, meetings and dinners to get principals on board were organized and held beyond the regular school day. This strategy allowed principals to meet each other and discuss approaches that their schools were taking with regards to state and district standards, problem-based learning, and other STEM activities. This setting also allowed time for principals to give critical feedback to CERC and other partners about the quality of the schoolwide professional development effort. The growth of the CERC professional development model credited how it was situated within the network of principals, which led to effective and sustainable STEM professional development.

Seeking principal involvement was significant for another reason. The ChildrenFirst Initiative gave decision-making authority to principals in the 1,700 schools, serving 1.1 million New York students. The principals, in turn, are now wholly responsible for educational programs within their schools (aligned to the standards for enhanced student performance), choices of partnerships and other support mechanisms outside the school, staffing and budget development, oversight and fiscal integrity. Principals are at the forefront of educational reform in the NYC public schools, based on the three-tiered approach of leadership, empowerment, and accountability spelled out by the ChildrenFirst Initiative.

To fulfill their responsibility, principals obtain support through a self-affiliation to over 60 ChildrenFirst Networks, each made up of 26–30 schools. These networks are designed to integrate operational and instructional support among schools. Each network uses a small cross-functional team to deliver personalized service to schools with the ultimate goal being to streamline operations and build capacity within each school so school-based staff can focus on instruction and accelerate student achievement. Self-affiliation is based on a number of factors determined by the principals, including common priorities, similar grade levels, geographic location, similar student demographics, and shared education beliefs or philosophies. Hence, it only seemed rational that principals would be critical to the development and implementation of change where the school was recognized as the unit of change.

Transition Point 1

Applying the Lesson Learned: Partnering With the Principal and Faculty of MS 88, Brooklyn, New York

CERC transition point 1 was characterized by its partnership with the principal and assistant principals and faculty of MS 88, a large Title I and Title III comprehensive school in NYC. The partnership was conceived with the essential objective of supporting core disciplines through an integrated STEM curriculum (I-STEM) design, development, and implementation, contextualized within project-based learning. The goal was to enhance a teacher-centered model by layering student-centered learning into a constructivist approach to teaching and improving student achievement. CERC developed I-STEM, which is a process characterized by using curriculum "connectors" called concepts along with imbedded processes guided by inquiry, that is, questions. Furthermore, I-STEM "comes alive" through experiential and project-based learning to help students acquire content knowledge and skills.

Through I-STEM, CERC focused on 15 concepts (e.g., change, nature, structure) that served as the glue for interdisciplinary, crosscutting themes, and multiple connections (Figure 12.3).

Figure 12.3

I-STEM CORE CONCEPTS IMPLEMENTED AT MS 88

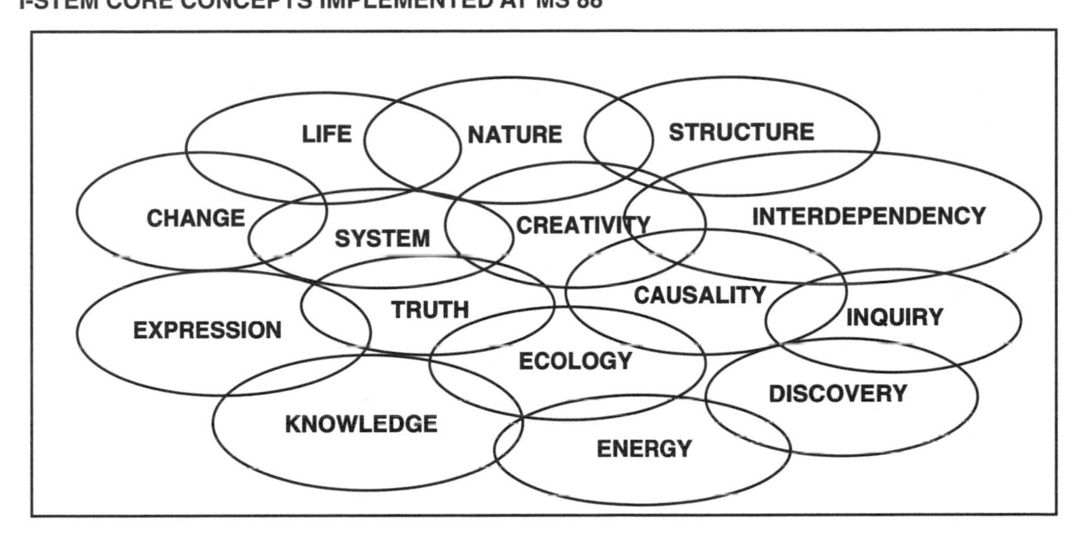

These concepts were selected because they had maximal appeal to teachers in all disciplines. Because concepts were defined as timeless, sharing common attributes, depicting emotional attraction, possessing intellectual power, appearing universal in scope, and embodied in *one* word, all teachers could "connect" to the concept and then bring in their big ideas about how concepts and disciplinary-based units of study would fit together. Often, faculty would connect over more than one concept, for example ecology and change.

As an example, under this construct, ecology was used as a concept to mean "the study of home" rather than the science discipline content. So as CERC's partners conducted their curriculum discussion and planning around the concept of ecology as home and change as process and measurement, a series of broadly based questions helped concretize their engagement with each other around their curriculum contributions.

The following questions were posed by real-world faculty interdisciplinary teams as they worked with the core curriculum linked through ecology and change: What is a home? How do we change our homes and habitat? How do our homes and habitat change us? Is change good or bad or both? What is common to home and habitat? What changes? How do we perceive, record, and measure change? What is common to every sort of change? What is the impact of change on people and the environment? What does change mean in science, in math, in literacy? How do science, math, and literacy help us understand change? How do these disciplines help us study our home?

The critical point here was that I-STEM was not a substitute for the existing curriculum. Rather, it drew from what was taught and when topics were covered during the academic year to better align them across subject areas and to reinforce major concepts and skills as students moved from science to mathematics classes. For example, one concept might be carried across subjects by explaining the powers of 10 using the reproductive capacities (change) of bacterial colonies (their home). In the literacy context, bacterial reproduction could be a topic for nonfiction writing, while other faculty members could work across disciplines to reinforce observation, inference, and critical-analysis skills.

Introduction of project-based learning: Integrated Project Week

To approach curriculum design through interdisciplinary or multidisciplinary engagement anchored in concepts and questions can be elusive and challenging. Project-based learning design, development, use, and assessment is a very important tool to make I-STEM real and intellectually, educationally satisfying. The Integrated Project Week (IPW) is the tool that CERC and MS 88 used to achieve these objectives. IPW is a capstone problem-based learning experience that takes place toward the end of each semester. IPW is designated over a period of 12 school days, with a total of 90 hours of full project-based activity embedded in the school year. IPW offers a perfect opportunity to build in-depth knowledge and skills and educational activity concretized through problem-based learning and characterized by 15 student-centered learning goals created by MS 88 faculty.

I-STEM IPW Student-Centered Learning Goals

The following 15 student-centered learning goals, developed by faculty, serve as benchmarks for I-STEM IPW curriculum formative assessment rubrics:

1. Identifies, articulates, synthesizes, and integrates important concepts, questions, and knowledge from a variety of disciplines.
2. Understands that knowledge is dynamic and cumulative and is valuable to daily life
3. Appreciates and understands the impact of human activity in defining and shaping culture and the natural environment
4. Demonstrates independent, critical, and creative thinking in developing hypotheses and theories
5. Gathers and critically evaluates sources of information
6. Examines, analyzes, and interprets information and data from multiple sources
7. Uses appropriate strategies in identifying and solving viable conflicts and problems
8. Communicates effectively and appropriately to a variety of audiences using an array of approaches
9. Sets realistic, challenging goals and objectives based on learning experiences.
10. Reflects on the learning process and offers and accepts critical feedback in a positive way

11. Encourages learning through collaboration with peers and mentors
12. Recognizes that emotional maturity, self-awareness, and integrity are important aspects of intellectual growth
13. Empathizes with the human condition
14. Recognizes the value of the global environment and ecosystem services.
15. Engages in activities to help ensure the continued welfare of their communities from an environmental, social, or economic perspective

Groups of students are assigned to an IPW course to explore, in an in-depth way, a current topic or theme linked to the inquiry processes. Although topics vary, all projects meet the New York State learning standards. The curriculum unit is a combination of classroom work, field trips, guest speakers, and workshops, through which students must complete concrete deliverables.

IPW projects cross disciplinary boundaries, bring together various aspects of the curriculum into meaningful associations, meet external learning standards and STEM integrated program goals and objectives, focus on conceptual areas of study, reflect the interests and needs of students, and display a constructivist and interactive pedagogy. The IPW is a crucial venue for helping to create deeper understandings and connections that might not otherwise emerge in the absence of the opportunity to spend an extended period of time on one project area.

From explorations and field and lab work throughout IPW, teachers also support students to construct STEM knowledge and discover meaning and significance for themselves by the questions they pose. The IPW closely connects students to each other within their own school as well as to the outside community. Finally, through the experience of IPW, the children begin to build a repertoire of intellectual, emotional, and social skills they will need throughout life. They learn about cultivating tolerance and self-respect and acquire a keener sense of caring and responsibility for humanity and the earth. This is all linked back to the core values established by the middle school faculty.

One of the most important aspects of IPW is that NYC is used as the living laboratory where students can examine both the human impact on the natural environment and the human built environment for both kinesthetic and visual learning (see Table 12.1, p. 193).

The deliverables for IPW may be presented in a variety of formats that include display boards and blogging, e-journals, video documentaries, scientific posters, reflections, and other active learning mechanisms. At the end of the IPW, a showcase celebration brings the entire community, including custodians, security and support personnel, faculty, students, parents, and guests together. Students and teachers visit each exhibit, performance, or presentation. The entire community is engaged to praise, document, and assess the creativity and skills of students.

Table 12.1 MS 88 IPW course description: Concepts, ecology, and change

Course number, title, and description	Teacher and grade	
Course 6001: CREATE YOUR OWN 'HOOD Have you ever wanted to build your ideal neighborhood? This is your chance. Kids will design every aspect of their home and neighborhood, including housing, streets, parks, recreation areas, and shops. We will pay close attention to our environment and see how the city and nature connect. Students will explore areas of Manhattan and Staten Island to get their creative juices flowing. Field trips to Historic Richmondton in Staten Island and Battery Park City in Manhattan.	Roumbeas and Rosenzweig	6
Course 6002: FASHION, CHANGE, AND ENVIRONMENT In this project, students explore how and what people and animals "wear" and how fashion is affected by the natural environment. By learning about different climates and how plants and animals adapt to them, kids will explore how and why people choose different fashions because of climate. All of this is in preparation for a global fashion show of student-designed clothing. Field trips are planned to Prospect Park Zoo, the American Museum of Natural History, and the Costume Hall of the MET.	Kim and Rios	6
Course 6003: ZOOKEEPERS NEEDED! By studying different kinds of ecosystems, students will transform their classroom into Park Slope Petland. Fish, reptiles, and plants will inhabit the habitats the kids create to simulate a desert, a lake, and a rain forest. Field trips to the Bronx Zoo and New York Aquarium are planned.	Bradley and Sullivan	6
Course 6004: BUY ME, BUY ME, GIMME, GIMME! Artists wanted! Kids will examine the work of the artist Edward Burtynsky and explore the relationship of industry and nature. Also, everyone needs to summon their most creative ideas to explore and expose the truth behind what you own today and what happens to it when you throw it away tomorrow. Original works of art will be exhibited at the MS 88 Showcase. Fieldtrip to the Brooklyn Museum of Art.	Capalbo and Lazzaro	6
Course 6010: NOISE POLLUTION Sounds are all around us. Some are pleasant and make us feel good, while others are just simply annoying. And when it comes to pollution, it's not just dirty air and bad water that we need to consider. Noise pollution, especially in urban areas, is a real problem. Students will learn about noise and how to measure sound and discover how an environment can be changed by decreasing noise pollution. To measure sound levels and degrees of noise pollution, students will venture out to the streets of Brooklyn and Manhattan to do hands-on scientific research. A field trip to Lincoln Center is planned.	Marczika and Rose	8
Course 6007: GUIDE TO A HEALTHY LIFESTYLE IN THE CITY This class is an exercise in healthy living. Students will develop a daily exercise program as well as a healthy—and tasty—diet to develop a truly "healthy lifestyle." To help other kids get and stay healthy, students will produce an exercise video. Field trip to Bally's Fitness Center.	Glaser and Mangiaracina	7
Course 6011: A BLUEPRINT FOR MY LIFE Who knows what life may bring? But that doesn't mean you cannot dream and plan to realize your aspirations. What does it take to make it in the NBA; win a Grammy; or become a scientist, mathematician, or computer programmer? Whatever goals you set, you will need a plan! Yet, everyone needs to be able to adapt and see all the options. This class will provide you with your own blueprint for life but will show you how to adapt when life throws you a curveball. Field trip to Columbia University and the Upper West Side of Manhattan.	Jasmin and Zangrilli	8
Course 6014: WILDLIFE IN THE CITY Students will explore how wildlife has changed because of the urban environment that we share with them. By investigating the coexistence of species, kids will document species adaptation through digital photography. Field trips include Prospect Park Zoo, Botanical Gardens, Brooklyn Center for Urban Environment, and Greenwood Cemetery.	Wolosky, Ortiz, and Medina	8

The outcome of CERC's work with MS 88 is noteworthy. When CERC and MS 88 began their partner work together, MS 88 was a school under registration review, which meant that the school was in danger of being closed. Now MS 88 has a rating of A and has remained that status for six years, even with a student body of nearly 1,000 middle school children. The principal continues to use the central components of the core curriculum integration and inquiry through I-STEM and IPW. Under the principal's leadership, education tools of excellent teaching and learning expanded to include differentiated instruction and the use of multiple assessment measures.

Transition Point 2

Throughout the first transition phase, to say that both the CERC and MS 88 staff and faculty were on a steep learning curve is an understatement. During the initial 24 months of engagement, CERC staff was on-site at MS 88 three out of five days a week, working with a designated assistant principal and teams of teachers. The teachers, Columbia University instructors, and CERC staff also met on a bimonthly basis on Saturdays to learn about urban ecology, fieldwork, and I-STEM, as well as to design, develop, and implement IPW.

Additionally, the principal of MS 88 and her leadership team advanced the logistics of curriculum delivery within the school schedule and organizational structure. This included providing planning and preparation time for and by faculty, adhering to the needs of education plans for special needs students (including English language learners), accounting for the regulations of the United Federation of Teachers, as well as addressing implications for the budget, other school partnerships, and, perhaps most importantly, the annual performance review required of all schools within NYC.

The partners grappled with both formative and summative assessment, specifically how to link IPW with student performance on standardized test (which is an ongoing process for both CERC and the schools and faculty). Because the faculty felt that it was engaged in a learning process as much as the students were, four sets of formative assessment rubrics were created: Student Self-Assessment, Student Technology Self-Assessment, Teacher Self-Assessment, and Teacher Technology Self-Assessment. Additionally, a rubric to assess knowledge and skills acquisition was also designed and used, with four areas comprising approximately 10 ranked questions. The areas were interactive learning, conceptual and experiential learning, final presentation and showcase, and personal character and behavior. Rankings range from 0 to 3, including did not demonstrate, needs improvement, proficient, and exemplary.

The entire student and teaching body of MS 88 completed the rubrics, and aggregate data was shared among the teaching staff. The principal created a grade for the IPW on all student report cards, and these grades are considered in the overall grade point average. The principal has also been instrumental in influencing the use of the I-STEM curriculum and IPW in several other schools in her ChildrenFirst Network.

Teacher self-assessments revealed another very important lesson for CERC: the realization that many science, mathematics, and literacy teachers had themselves never been through a hands-on inquiry process nor had they been exposed to in-depth project-based learning.

So, CERC's initial thoughts that all science and math teachers had a solid handle on the pedagogy of inquiry shifted. In order to teach inquiry really well, one had to experience the process. Additionally, the science content had to be active, robust, intellectually engaging, challenging, and memorable. In effect, all the characteristics CERC wanted to see in the experience for middle school children also had to be experienced by participating teachers.

To meet these higher needs for teacher knowledge and skills, CERC had to improve its approach to professional development for teachers. To do this, CERC turned to three sources: (1) the SEE-U program; (2) the NSES for teacher professional development; and (3) its partners in the CERC consortium, the Lamont Doherty Earth Observatory, and the NYC Department of Education. With respect to point 3, over a three-month period CERC engaged in a series of meetings and discussions with scientists and researchers from the Museum of Natural History; the New York Botanical Garden; the Wildlife Conservation Society; and divisions of the Lamont Doherty Earth Observatory, including Biogeoscience, Geochemistry, and Climate Science. Professional staff from the NYC Department of Education's division of math and science were also part of the faculty meetings. These meetings focused on how we could improve the content of professional development linked to the living laboratory that is NYC and deliver rich, robust science through place-based learning. For instance, the marshes and wetlands of NYC could be revealed through the study of native and invasive plants, island ecology, changes in landscape and ecosystems characterized by geology and biodiversity. These discussions were then concretized in lectures and hands-on learning to enhance professional development of teachers at MS 88.

Concurrently, CERC felt that it had moved beyond the beta-testing of the work with MS 88 and was now ready to apply for external funding for which it would propose to combine an intensive summer fieldwork and academic year lectures, seminars, and workshop with its in-school implementation to ensure that professional development would take root. For CERC, the one *overarching goal of "taking root" would be realized when: (1) the middle school core curriculum was fully imbued with I-STEM and project-based learning through IPW; (2) I-STEM and IPW actually became an integral part of middle school core curriculum design, delivery, and assessment on an ongoing, long-term basis; and (3) when CERC's faculty and researchers were able to consistently design and deliver the highest level of science teaching and experiential learning in the professional development.

Transition Point 3

To realize the higher standard required of first-rate professional development designed to support teachers in engaging their students in active learning, CERC applied for and was awarded an National Science Foundation Innovative Technology Experiences for Students and Teachers (ITEST) grant called Technology, Research, Ecology Exchange for Students, or TREES. The TREES project is presently the professional development model for our work. It brings together important aspects that CERC believes can provide the type of teaching and learning that will move teachers and students forward in STEM and STEM ICT education (Figure 12.4).

Figure 12.4

CURRENT MODEL OF I-STEM AND STEM ICT PROFESSIONAL DEVELOPMENT

I-TEST = Innovative Technology Experiences for Students and Teachers; *NGSS* = *Next Generation Science Standards.*

On the basis of its prior work, CERC identified three key elements for the third transition of its professional development model, which included: (1) continuing to work with teachers to enhance the core curriculum of the middle school through an interdisciplinary framework in which STEM, STEM ICT, and IPW are emphasized; (2) continuing to use university-level field-methods courses to inform the quality and rigor of professional development content and skills for inservice middle school teachers; and, (3) framing points 1 and 2 through the appropriate clusters of standards, including the *Next Generation Science Standards.*

A final central point to transition point 3 is that TREES marked the expansion from MS 88 to five other large Title I and Title III schools, each with its own mission statement and faculty visions, with TREES core values agreed to by all as foundational for the learning community of middle school faculty who attend the TREES program (Table 12.2). The core values include respect, honesty, commitment, empowerment, and balance.

Table 12.2 Characteristics of schools participating in the TREES professional development

TREES project organizational partners								
(School statistics 2010–2011 school year)								
School	**Type**	**District (D)**	**Grades**	**Enrollment**	**Class size**	**Title I**	**Demographics**	**Math or science overall accountability**
MS 331: Bronx School of Science Inquiry and Investigation	Urban	NYC D 10	6–8	371	29	Yes	1% White; 30% African American; 68% Hispanic *Poverty rate 96%*	Good standing
MS 324: Patria Mirabal	Urban	NYC D 6	6–8	428	28	Yes	1% White; 13% African American; 86% Hispanic, 1% American Indian *Poverty rate 93.8%*	Good standing
IS195: Roberto Clemente	Urban	NYC D 5	6–8	552	28	Yes	2% White; 1% Asian, 48% African American; 49% Hispanic; 1% American Indian or Native Alaska *Poverty rate 88.8%*	Good standing
JHS 088: Peter Rouget	Urban	NYC D 15	6–8	891	24	Yes	11% White; 15% Asian; 16% African American; 56% Hispanic; 1% American Indian or Native Alaska *Poverty rate 80.5%*	Good standing
JHS 052: Inwood	Urban	NYC D 6	6–8	728	32	Yes	1% White; 1% Asian; 3% Black; 96% Hispanic; 1% American Indian or Native Alaska *Poverty rate 93.9%*	Good standing

Source: New York City Department of Education. *http://schools.nyc.gov/default*

Building on the foundational bedrock of the 1996 NSES and reflecting on the future, CERC has been exploring ways to incorporate the *Next Generation Science Standards* (*NGSS*; NGSS Lead States 2013) goals and *Common Core State Standards* (NGAC and CCSSO 2010) into professional development in a cohesive and cogent way. Because teachers work with concepts from I-STEM and interdisciplinary planning, crosscutting concepts, core ideas, and practices from *A Framework for K–12 Science Education* (NRC 2012) and the *NGSS* are familiar. TREES faculty have done the work of identifying synergies between and among the *Framework*, the *NGSS*, and the *Common Core State Standards*, with specific reference to units of study in their core curriculum. While they have started with units that line up easily, they have gone through the necessary processes so that more difficult units—such as those of evolution and natural selection—may be more readily analyzed and aligned.

CERC also encourages TREES teachers to collaborate and connect with other teachers in their change process because they believe that teachers learn better when they exchange ideas about ways to integrate crosscutting themes into their teaching. Because teachers participating in TREES have shared goals aimed toward mutual support and reinforcement of each other and their students, expertise among the TREES cohorts is revealed and shared. In short, the teachers help each other and experience satisfaction in doing so. Moreover, recognizing that the history, culture, and organization of each school is unique, CERC uses continuous program assessment to ensure quality across all delivery components of the five schools involved in the partner effort. This assessment takes place in a number of ways, including frequent staff meetings to identify and solve problems, as well as an extensive, confidential program survey that is completed by TREES teachers at the end of each TREES Institute. Survey data are compiled and analyzed by a third party, and results are conveyed to the TREES principal investigator and CERC staff. Finally, the principal investigator also meets with principals and assistant principals from TREES schools four times during the academic year to solicit feedback and insights on what is and is not working. This information is then woven back into design and implementation.

Through TREES, teachers engage in the intellectual exercise and experience of fieldwork in urban ecosystems, research, and scientific presentation as well as case study analysis and role-playing on a topic of STEM or sustainability science significance. The goal here is to support the teachers in being students again, not only for their knowledge content and skills enhancement but also to mirror what their own students might experience in inquiry and experiential learning.

To build the teacher's content knowledge about ecology and urban ecosystems, a two-week professional development program is offered. The first week is devoted to a series of field trips to wetlands and marshes. In the marshes, CERC scientists expose teachers to the practical elements of field science. The marshes provide an excellent locale for study of ecology as well as a setting in which extensive interaction occurs between humans and the environment. The program is designed to immerse the teacher participants in the subject matter. In order to do so, teacher participants collect data and conduct fieldwork at Inwood Park and Piermont Marshes in New York City.

On the basis of these field activities, teams of teachers work on various science research projects to explore marsh plant and animal species as well as other characteristics and trends, including salinity, turbidity, pH balance, and oxygen levels. The teacher groups present their findings or propose research projects during sessions for the faculty and peers. Additionally, CERC is working on ways to help teachers integrate technology into their classrooms by providing technological knowledge and tools. These tools include computer probes and software packages for experiments, data analysis, and data representation.

The second week of the program introduces the teacher participants to project-based learning using *Understanding by Design* (Wiggins and McTighe 2005). In interdisciplinary teams, teacher-participants also learn about how to develop an IPW curriculum for later use with their schools and students by "planning backwards."

Also, part of the second week is devoted to a case study of the Marcellus Shale hydraulic fracturing, in which teachers evaluate the various political, social, economic, and environmental elements involved in the decision making about hydraulic fracturing in New York State. The discussion focuses on weighing the economic and social benefits of hydraulic fracturing, such as job creation, versus the potential environmental damage hydraulic fracturing could cause. In conjunction with time spent studying hydraulic fracturing, the teachers also work in teams to produce samples of potential activities and topics for IPW. The projects expand on the first week's marshes and wetlands lectures and field trips and expound on the case study, all while developing the IPW to be integrated into the curriculum requirements for the upcoming year.

Finally, for those who may want to use field ecology courses as a basis for professional development, it is useful to describe salient parts of the adaptation process. As indicated previously, CERC has used a five-week program, the SEE-U program, as the basis for the study of the NYC ecosystems upon which TREES is designed. SEE-U students learn essential elements of the scientific and fieldwork method through island ecology and ecosystem science content in field sites in South America, the Caribbean, and the Middle East. During the five weeks, students work on four principle areas: (1) biomes, biotic processes, abiotic processes and contemporary issues through modules including lectures and guest speakers; (2) individual research projects guided by a testable question specific to the field site—projects involve a review of relevant literature, a final research proposal, field data collection, data analysis and a final presentation; (3) field practical during which, as teams, students present their ideas, discuss how they will collect and analyze field data, and give a final oral presentation; and (4) journaling and blogging for which each student keeps a notebook journal while at field sites to record work both in the field and in lecture. Entries are required to reflect the individual's experience at the site and day-to-day participation in online and seminar-based discussions on the ethics of conservation. Finally, nine key skills acquisitions complete the program (Table 12.3, p. 201).

Figure 12.5

ACQUISITION OF NINE KEY SKILLS

1. Knowledge and use of scientific methods used in experimental design	4. Integration of data from many sources and placement into correct theoretical context	7. Ability to conduct appropriate statistical tests
2. Independent gathering of data from nature and geo-referencing sources	5. Understanding of Geographic Information Systems (GIS) and Global Positioning System (GPS)	8. Presentation of cogent, engaging oral and written scientific research reports
3. Differentiation between relevant information and data that are interesting but irrelevant	6. Manipulation and understanding of various computer models	9. Ability to reflect upon research work as an iterative and ongoing process

The adaptation of SEE-U to the TREES project met with constraints on time and geography because the project team found that teachers could not participate in the five consecutive weeks of instruction included in the original proposal even though the professional development would take place solely in urban ecosystems in NYC. To accommodate these constraints and with any research experience needing to be composed to a significant level of guided inquiry, CERC decided to focus on real-world research by scientists from Columbia University already underway in local urban ecosystems in the marshes, waterways, and urban forests. In this manner, teachers could experience what it is like to observe and collect data and then augment it with the existing data of the research protocol.

Reflections for the Future

Assessment Linked to Testing Is an Important Next Step

The importance of TREES as a professional development model, in the final analysis, rests in its potential contribution to improving the lives of our children and young people through powerful STEM and STEM ICT education. This means that linking professional development to testing and performance is crucial.

Although NYC's four-year high school graduation rates have been increasing in recent years, ratings in 2008 were still about 10% below the state (70%) and national (74.9%) averages (Rampell 2009; NYCDE 2009). Racial differences also play a role in determining educational success, with 61.5% of black and 63.5% of Hispanic students graduating nationally, compared to 81% of white students (Rampell 2009). Public school students in New York State are also presented with the additional challenge of passing the mandatory Regents High School Exams to receive their diploma. In light of these statistics, it is imperative that we develop strategies to help prevent our students from falling short of state and national averages in the future.

Low graduation rates are attributed to several key factors, none of which is mutually exclusive. Often the blame is placed on the student for lack of motivation in school, deviant behavior, poor attendance, or the influence of family and socioeconomic factors (Battin-Pearson et al. 2000). Poorly funded schools, ineffective curriculum, and lack of opportunities for school bonding are other institutional challenges that can lower graduation percentage rates (Battin-Pearson et al. 2000; Fleischman and Heppen 2009).

Middle school (typically grades 6–8) is a pivotal time for preparing for the transition to high school and for determining if a student will complete four years (Neild 2009). Middle school students are also exposed to the Regents Exam in grades 6 and 7 for English language arts and mathematics, with the addition of science and social studies in grade 8. This component of the Regents Exam prepares students in terms of format and practice for their graduating exam, and it is also a determining factor in a student moving on to the next grade (No Child Left Behind 2001). In an attempt to meld standardized testing expectations with improving student achievement and engagement in the classroom, professional development for public middle school teachers became the focus of this analysis as we work under the premise that it is the teacher who is the agent and/or target of change and who has the most impact on student achievement.

CERC Approach to Advance Student Performance

CERC's partnership with middle schools helps to improve teaching and learning and is aimed at battling some of the fundamental causes of low student achievement and high dropout. As noted earlier, the impetus to move to the middle school level was to revert the impact of poverty on student learning. Hence, CERC sought to improve student interest and retention in school through teacher professional development and through engaging students in pedagogical instructional methods that might increase their interest in STEM and STEM ICT. Since evidence from SEE-U and MS 88 showed the CERC methods to be effective, the intent is to use lessons learned to move these activity-based learning experiences to the middle grades. Whole-school involvement can use an integrated curriculum that focuses on central themes that can be connected across all subjects, while project and problem-based learning are conducted over a shorter period of time and are more focused on specific skills or content (Schneider et al. 2002).

Using Standardized Testing as a Tool

Despite widespread shifts in educational opportunities, standardized testing, which is deeply embedded in politics, remains a fixture in the New York State Department of Education as a means of assessing a student's academic performance (McDonnell 2005). Often, standardized testing has a negative connotation because it narrows what content is taught in the classrooms, and schools are judged, and sometimes funded, on the quality of student's scores (Moon et al. 2007).

CERC is preparing teachers to use standardized testing as a tool, through using activities that ensure that students are prepared in content, familiar with the exam format, and have the ability to apply more abstract, real-world learning to testing situations. From this perspective, inquiry-based learning is viewed as a different approach to actually helping students deal with what they will encounter on a standardized test.

Looking Forward: Measuring Student Achievement

To connect project and problem-based learning with improving standardized testing, the next step is to use the assessment tools in the classroom. To do this, teachers are taught how to incorporate the testing assessment in a positive manner such that students would be proud of their ability to apply what they learned to the novel situations, such as assessment tests, that eventually count toward their graduation success. Because project- and problem-based learning evaluations are typically group orientated, CERC is suggesting that teachers begin with group or partnered testing and then move toward individual grading when students are more comfortable with the format. Ideas for introducing testing components include quiz shows or asking students to write their own questions based on what they learned during the project unit.

Released questions from previous tests could be used as a guide rather than a strict requirement, thus serving to replace or supplement poorly quantifiable rubrics. CERC found this to be especially useful for English language arts students, for which the most commonly used phrasing of the question on the Regents Exam can be used in multiple contexts throughout the project in the form of essay questions and listening comprehension. Teacher creativity in presenting the questions and assessments is key.

Through professional development activities, CERC introduces tools that teachers can implement in their classrooms during a trial-and-error period that may help teachers see how students are responding and record achievement levels on exam-type questions. Trend data from such exercises should be analyzed to understand where the weaknesses and strengths are and if students are scoring higher on project- and problem-associated exams compared with students using regular lesson plans. It will also be important to assess where content is lacking or advanced in developing curriculum compared with testing requirements.

Conclusion

Learning to make changes in services offered to schools is key to improving the performance of teachers and students. When CERC realized that its teacher-centered professional development was not effective, CERC changed to a different model to hone in on gaps in what was being offered. However, when CERC found that even the move to inquiry-based teaching fell short of expectations, it again realized that it had made assumptions about teachers' ability to implement inquiry without finding out if teachers know how to teach using that instrumental format. That said, CERC took the change one step further by reverting back to a program proven to be successful with undergraduate students and adapting it to the middle schools while pairing it with concepts relevant to students' lives. CERC then also linked this to existing ongoing scientific research by scientists from Columbia University. This final approach appears to be a model worth the investment of time and effort by a host of partners.

CERC saw the opportunity to use assessment tools in combination with project and problem-based learning as another way to provide teachers and students with a quantifiable performance measure and aid in Regents Exam preparation. The core of what makes the STEM activities successful (i.e., integrated curriculum learning, hands-on activities, generating enthusiasm in the subject matter) should not be minimized. Active learning about STEM topics relevant to students' lives enables them to take knowledge and skills gained and apply them to new situations and comparable questions they will face in the future. This will in turn enable NYC teachers and students involved in CERC projects to think beyond the immediate lessons to ways to use these tools for success that can be of benefit to them both academically and professionally.

References

Acemoglu, D., and J. Angrist. 2000. How large are human capital externalities? Evidence from compulsory schooling laws. *NBER Macroeconomics Annual* 15: 9–59.

Battin-Pearson, S., M. D. Newcomb, R. D. Abbott, K. G. Hill, R. F. Catalano, and J. D. Hawkins. 2000. Predictors of early high school dropout: A test of five theories. *Journal of Educational Psychology* 92 (3): 568–582.

Bridgeland, J., J. Dilulio, and K. B. Morison. 2006. The silent epidemic: Perspectives of high school dropouts. Bill and Melinda Gates Foundation. *Gatesfoundation.org/nr/downloads/ed/TheSilentEpidemic3-06FINAL.pdf.*

Drake, S. M., and R. C. Burns. 2004. *Meeting standards through integrated curriculum.* Alexandria, VA: ASCD.

Farstrup, A. E., and S. J. Samuels, eds. 2002. *What the research has to say about reading instruction.* 3rd ed. Newark, DE: International Reading Association.

Fleischman, S., and J. Heppen. 2009. Improving low-performing high schools: Searching for evidence of promise. *Future of Children* 19 (1): 105–133.

Kurlaender, M., S. F. Reardon, and J. Jackson. 2008. *Middle school predictors of high school achievement in three California school districts.* Santa Barbara, CA: California Dropout Research Project. *cdrp.ucsb.edu/dropouts/pubs_reports.htm.*

McDonnell, L. M. 2005. No Child Left Behind and the federal role in education: Evolution or revolution? *Peabody Journal of Education* 80 (2): 19–38.

Meece, J., and J. S. Eccles, eds. 2008. *Schooling and development: Theory, methods and applications.* Hillsdale, NJ: Erlbaum.

Moon, T. R., C. M. Brighton, J. M. Jarvis, and C. J. Hall. 2007. *State standardized testing programs: Their effects on teachers and students.* Storrs, CT: The National Research Center on the Gifted and Talented, University of Connecticut.

National Governors Association Center for Best Practices and Council of Chief State School Officers (NGAC and CCSSO). 2010. *Common core state standards.* Washington, DC: NGAC and CCSSO.

National Research Council (NRC). 1996. *National science education standards.* Washington, DC: National Academies Press.

National Research Council (NRC). 2012. *A framework for K–12 science education: Practices, crosscutting concepts, and core ideas.* Washington, DC: National Academies Press.

Neild, R. C. 2009. Falling off track during the transition to high school: What we know and what can be done. *Future of Children* 19 (1): 53–76.

New York City Department of Education (NYCDE). 2009. School accountability tools. *schools.nyc.gov/Accountability/tools/default.htm.*

NGSS Lead States. 2013. *Next Generation Science Standards: For states, by states.* Washington, DC: National Academies Press. *www.nextgenscience.org/next-generation-science-standards.*

Rampell, C. "SAT scores and family income," *Economix: Explaining the science of everyday life* (blog), *New York Times*, August 27, 2009, *economix.blogs.nytimes.com/2009/08/27/sat-scores-and-family-income.*

Roediger, H. L., and E. J. Marsh. 2005. The positive and negative consequences of multiple-choice testing. *Journal of Experimental Psychology* 31 (5): 1155–1159.

Rumberger, R. W. 1983. Dropping out of high school: The influence of race, sex, and family background. *American Educational Research Journal* 20 (2): 199–220.

Schneider, R. M., J. Krajcik, R. W. Marx, and E. Soloway. 2002. Performance of students in project-based science classrooms on a national measure of student achievement. *Journal of Research in Science Teaching* 39 (5): 410–422.

Wiggins, G., and J. McTighe. 2005. *Understanding by design.* Alexandria, VA: ASCD.

Part III

Studying Models and Approaches to STEM Professional Development in a Professional Learning Community Setting

Chapter 13

Creating and Sustaining Professional Learning Communities

Jane B. Huffman

Editors' note: The editors believe the chapters within this book will be a valuable source of information and study for those schools, districts, and states that are working to enhance or reformulate their science, technology, engineering, and mathematics (STEM) professional development models. We also believe that an extremely effective means for the study of current models and the thoughtful planning of new, more effective models is through the use of professional learning communities. To that end, we invited one of the foremost experts in our country in implementing professional learning communities as a basis for educational improvements, Dr. Jane Huffman, to contribute to this volume. We know that you will find her insights and the information contained in this chapter helpful as you use this book to analyze your own professional development models and implement new strategies for continuously improving the effectiveness of STEM professional development at all levels. —*Brenda S. Wojnowski and Celestine H. Pea*

After a decade of working with schools to provide processes and strategies related to school improvement and student achievement, the list of lessons learned could fill a book. For this chapter, however, the emphasis is to highlight the most important considerations when attempting to move a school toward the creation of an effective learning community. Those considerations are (1) shared values, vision, and leadership; (2) collaborative teaching, learning, and application; and (3) professional and personal efficacy and commitment. These are essential to the professional learning community (PLC) process, and in assuring that the process and strategies will be sustained. Embedding these considerations in a school or district requires that it be recultured, not merely restructured. As Fullan (2006) notes, "[PLC] is a cultural change that is both deep and necessary." Through our work, we have come to know the challenges these words pose.

What follows is a summary of considerations related to the establishment of the PLC process. This includes quotes from practitioners in schools that are implementing PLCs (Hipp et al. 2008). I close with seven lessons learned from the schools that participated in our recent research.

Shared Values, Vision, and Leadership

Changing the culture of an organization is a difficult and time-consuming process that must embrace a vision shared by all stakeholders (Huffman 2003). Yet simply announcing the vision and imposing it on staff members will not develop the energy and commitment needed to implement substantive change. The task of the leader is to encourage and enable faculty to combine their personal values and ideas into a collective vision, which becomes an integral component of the change process as it emerges over time. This collective vision provides direction to align curriculum, instruction, and assessment and also supports other school initiatives. By developing a vision based on shared values, the school culture honors the dedication and talents of individuals who seek to provide a supportive and successful environment for students. Here are the words of those school practitioners in our study who are making this vision synthesis happen:

> It doesn't happen overnight; it takes a few good leaders as well as teachers and administration to get the whole group going in a whole new direction. … Our school is really working toward getting everybody on the same page and tackling our problems.
>
> The goal is to produce a lifelong commitment to learning and success in both academic and technological fields as well as others. … Basically every decision that we make focuses on student learning.
>
> [Leadership in the school is] like an onion; it's in layers because we have so many different people heading up different areas.
>
> Our leadership team is made up of teachers who have been trained in facilitation. They come from all different levels and curriculum disciplines. … We see new faces taking on leadership roles.
>
> Central office is very supportive. This is what sets us apart from other districts. Our administration gets us what we need. This is the message passed down from our superintendent.

Collaborative Teaching, Learning, and Application

In every article, book, or research report on PLCs, collaboration is mentioned. The admonishment is always to deprivatize classroom practice and create a new norm of sharing what *is* done and what *could and should be* done to improve teaching for the benefit of students. The very use of the term PLC requires that work in schools is collaborative. Furthermore, it implies that teachers are continuously learning, that is, "Professional learning that increases educator effectiveness and results for all students occurs within learning communities committed to continuous improvement, collective responsibility, and goal alignment" (Learning Forward 2012).

Of the utmost importance as well is application. To know but not to do characterizes many a teacher and school administrator. If PLCs are to thrive, then it is incumbent on all professionals in the school to be continuous learners and to apply this learning to their planning and delivery of instruction. Several teachers and administrators in our study have remarked on this subject:

> We are working more on ourselves to improve student achievement. We search for new ideas, and hopefully our learning will filter down to the students, but that's going to take time. We always think of students first. It's amazing to watch how much people can learn just feeding off of each other.
>
> Teachers have good ideas and aren't hesitant to speak up. They provide suggestions, make decisions, ask other's opinions, and adjust accordingly.
>
> There are teachers who, no matter what goes on in their classroom, keep teaching and get out into the halls talking, sharing, and urging others on.
>
> A big step is the coaching and relying on peers, having someone to turn to for help, and sharing that—I want to call it teaching wisdom—and it's not only with veterans, but new teachers have a lot to share too.
>
> The PLC seems to be the heart of what we're trying to do here, and that is trying to learn new strategies, new ways to reach our standards, and then also to become a "cut above" school.

Professional and Personal Efficacy and Commitment

Have you ever heard the saying, "When the going gets tough, the tough get going"? In establishing an effective learning community, this needs to be the mantra of everyone involved. All changes—major or minor—be they a new schedule, new teaching materials, a change in instructional practices, or an entirely new model of professional development, will produce awkwardness, frustration, trepidation and a desire to retreat to the more familiar. In reculturing, these same feelings will surface again and again. Changing a culture happens over time—a long time. Things go wrong, they don't move as rapidly as hoped, and they are often in direct competition with other personal and professional demands placed on teachers and administrators. The temptation to "call it quits" will arise frequently and for good reasons. Thus, when you reculture, everyone in the school or education system must commit to a level of persistence that may never before have been expected of them. Professionals must find new confidence in themselves and in their colleagues. Collegial relationships will be needed more with this change than most any other the school, district, or state has tried. The professional and personal efficacy and commitment of everyone will be tested. It is the mutual support of the faculty and all educators that will reenergize the efforts during difficult times. Without supportive relationships and actions, individuals will falter and the group will turn to commiserating rather than growing professionally. To avoid this failing, professional development should focus on processes such as data analysis, the use of protocols to discuss student work and teaching practices, conflict management, coaching, problem solving, and team building. The focus must be on what each individual and the staff as a whole can contribute to solving the problems the work of the school presents. In the field, professionals describe this as follows:

> I think we stumbled across a little catch phrase, "We believe," and that pretty much says it in a nutshell. We believe, and you can put the three dots behind it. We believe in our students. … We believe in our school. … We believe our students can learn. … We believe they have the right to learn. … We believe we have the right to teach. … We believe in everything. … We believe that anything is possible.
>
> Teachers have the freedom to do what they think is best for students. The [administrators] don't tell me how to teach. I can change what needs to be changed to benefit students, to be fair to students.
>
> I think our administrators trust their teachers with their ideas and what they think about their students because, after all, we know our students. And they trust us, and it's mutual.
>
> I think everyone is seen as important—an important part of the puzzle, that everyone contributes regardless of your role or your title. And I think what everyone realizes is that what they do matters and counts.
>
> I just think it's a great school to work at. I love coming to school everyday. [We are] very fortunate to have some very qualified teachers that love what they're doing, and you can see it. The extra time that teachers and administrators put into what they're doing—I don't think there's a school that works harder than this school. I just feel good about coming and our students, uh, I love our students.

Lessons for Practitioners

So, what has all of this study, research, and conversation with practitioners taught us as PLC researchers about changing a school culture to create and sustain a PLC? Changes that go to the very heart of the organization, that demand the creation of new norms, generate not merely resistance from a few naysayers but can also strike fear in those eager to embrace new ways of operating. Any change brings the need to let go of one thing and embrace something else (Bridges 1997). It requires the confidence to not only step out to greet the new way but to do so with both feet—letting go of the current and welcoming the future, though it may not yet be clear what that entails. Understanding this is basic to benefiting from the lessons shared here.

It is hoped that the following seven lessons, from schools that are successfully implementing the PLC process, will provide a clearer way for those who choose to use the PLC model as a way to improve student learning.

1. First, without question, the teacher's focus must constantly be on students. It is clear that all teachers contribute to the success of all students and that everyone

is accountable for cumulative student progress. Schools that lose sight of student learning as their ultimate goal are likely to lose their way during this long and winding journey.

2. Next, everyone needs to be prepared to work on this change for years, not days or months. Reculturing takes time, distributed leadership, and organized effort. Although there is a beginning to this journey, there really is never an end. Everyone in the school must be committed to this work for "the long haul."
3. It is essential that leadership be shared at all levels. Leadership in a successful professional learning community is defined by behaviors and strengths, not by position or title. Teacher leaders strengthen the depth and breadth of leadership and provide consistency of action throughout the school or system.
4. There can no longer be private practice in isolated classrooms. The expected behavior is that teachers and administrators share their practice. Professional development that ensures teachers are comfortable in sharing their practices and supporting their colleagues must be embedded in the school day.
5. Supporting school culture and maintaining effective learning communities necessitates a transparent plan for ongoing efforts to nurture relationships, embrace alternative practices, and sustain the work that will maintain integrity amidst changing conditions and leaders.
6. Central office support and interaction is critical to initial and long-term success. Only recently has the literature acknowledged the important role that districts' leadership plays in supporting the creation and sustaining of PLCs. If the school is to truly reculture, district leadership must provide the example and support. Thus, the district must be ready to reculture as well (Huffman, Pankake, and Munoz 2007).
7. Finally, there must be an awareness and an acceptance that resistance will occur. School personnel must prepare, lead, negotiate, and make reasonable and inclusive decisions based on what's best for students. Resistance can be good as well as bad; resistance can prompt us to look more closely, to question, and to make better decisions.

Final Thoughts

A comprehensive understanding of each of these considerations is far more than can be included in this chapter. Those interested in working to develop effective STEM PLCs are encouraged to read and study the abundant literature on PLCs. There are numerous writings that include case studies of schools and districts that have taken this path—some successful, others disappointing.

For example, Supovitz and Christman (2003) provide a valuable source in honing craft knowledge that fosters *communities of instructional practice* in their study, "A Tale of Two Initiatives" in Cincinnati and Philadelphia. They found "evidence to suggest that those communities that did engage in structured, sustained, and supported instructional discussions, and that investigated the relationships between instructional practices and student work, produced significant gains in student learning" (p. 5). Nonetheless, the majority of their teams did not deepen their practice over time, which suggests why reforms often fail. There is much to be learned from the experiences of others. I have had the privilege of being part of both the successful and unsuccessful attempts to develop and sustain PLCs. I share our lessons in hopes that others' experiences may be successful and provide collaborative opportunities for continued student and teacher learning through reculturing their schools as professional learning communities.

References

Bridges, W. 1997. *Managing transitions: Making the most of change.* London: Nicholas Brealey.

Fullan, M. 2006. Leading professional learning. *School Administrator* 63 (10): 10–14.

Hipp, K. A., J. B. Huffman, A. M. Pankake, and D. F. Olivier. 2008. Case stories: Sustaining professional learning communities. *Journal of Educational Change* 9 (2): 173–195

Huffman, J. 2003. The role of shared values and vision in creating professional learning communities. *NASSP Bulletin* 87 (637): 21–34

Huffman, J. B., A. M. Pankake, and A. J. Munoz. 2007. The tri-level model in action: Site, district, and state plans for school accountability in increasing student success. *Journal of School Leadership* 16 (5): 569–582.

Learning Forward. 2012. *Standards for professional development.* Oxford, OH: Learning Forward. *learningforward.org/standards/learning-communities.*

Supovitz, J. A., and J. B. Christman. 2003. Developing communities of instructional practice: Lessons from Cincinnati and Philadelphia. Policy brief no. RB-39. New Brunswick, NJ: Rutgers University, Consortium for Policy Research in Education.

CONTRIBUTORS

Foreword
Patricia M. Shane, PhD
Educational Consultant
1289 N. Fordham Boulevard, #266
Chapel Hill, NC
pshane@unc.edu
919-201-1981

Chapter 1
Celestine H. Pea, PhD
Program Director
National Science Foundation
4201 Wilson Boulevard
Arlington, VA 22230
cpea5@cox.net
703-568-2749

Brenda S. Wojnowski, EdD
President and CEO
WAI Education Solutions
Wojnowski and Associates, Inc.
6318 Churchill Way
Dallas, TX
bwojnowski@gmail.com
214-288-9779

Chapter 2
Celestine H. Pea, PhD
Program Director
National Science Foundation
4201 Wilson Boulevard
Arlington, VA 22230
cpea5@cox.net
703-568-2749

Brenda S. Wojnowski, EdD
President
WAI Education Solutions
Wojnowski and Associates, Inc.
6318 Churchill Way
Dallas, TX
bwojnowski@gmail.com
214-288-9779

Chapter 3
Joseph Krajcik, PhD
Director, Institute for Collaborative Research for Education, Assessment and Teaching Environments for Science, Technology, Engineering and Mathematics (CREATE for STEM)
Professor, Science Education
Michigan State University
College of Natural Science
College of Education
115 Erickson Hall
East Lansing, MI 48824
krajcik@msu.edu
517-432-0816

Chapter 4
Richard A. Duschl, PhD
Waterbury Chair Professor
College of Education
Penn State University
University Park, PA 16802
rad19@psu.edu
814-867-2759

Chapter 5
Karen Charles, PhD
Education Research Analyst
RTI International
126 Friar Tuck Road
Statesville, NC 28625
kcharles@rti.org
919-541-8056

Chapter 6
Jane Butler Kahle, PhD
Condit Endowed Professor of Science Education, Emerita
Miami University
16313 Somerset Drive
Broomfield, CO 80023
kahlejb@miamioh.edu
303-954-8825

Sarah Beth Woodruff, PhD
Director
Ohio's Evaluation & Assessment Center for Mathematics and Science Education and *Discovery* Center
Miami University
210 E. Spring St.
408 McGuffey Hall
Oxford, OH 45056
woodrusb@miamioh.edu
513-529-1686

Chapter 7
Jeff C. Marshall, PhD
Director, Inquiry in Motion Institute
Associate Professor, Science Education
404A Tillman Hall
Clemson University
Clemson, SC 29634-0705
marsha9@clemson.edu

Michael J. Padilla, PhD
Director and Associate Dean, Eugene T. Moore School of Education

102 Tillman Hall
Clemson University
Clemson, SC 29634-0702
padilla@clemson.edu

Robert M. Horton, EdD
Professor, Mathematics Education
409B Tillman Hall
Clemson University
Clemson, SC 29634-0705
bhorton@clemson.edu

Chapter 8

Jean Cate, PhD
Associate Director for University
and Academic Partnerships
K20 Center
University of Oklahoma
3100 Monitor Avenue, Suite 200
Norman, OK
jcate@ou.edu
405-325-2228

Linda Atkinson, PhD
Associate Director for K12 STEM Partnerships
K20 Center
University of Oklahoma
3100 Monitor Avenue, Suite 200
Norman, OK 73072
latkinson@ou.edu
405-325-2228

Janis Slater, MEd
Science Coordinator
K20 Center
University of Oklahoma
3100 Monitor Avenue, Suite 200
Norman, OK
jslater@ou.edu
405-447-8000

Chapter 9

Katherine Hayden, EdD
Director of School of Education
Professor of Educational Technology
California State University San Marcos
333 So. Twin Oaks Valley Road
San Marcos, CA 92096-0001
khayden@csusm.edu
760-750-8545

Youwen Ouyang, PhD
Professor
Department of Computer Science
and Information Systems
California State University
333 South Twin Oaks Valley Road
San Marcos, CA 92096-0001
ouyang@csusm.edu
760-750-8047

Nancy Taylor
Executive Director
San Diego Science Alliance
5694 Mission Center Rd. #285
San Diego, CA 92108
nancy.taylor@sdsa.org
619.400.9777

Chapter 10

Sidney Smith
sidsmith8@gmail.com

Marilyn Decker
Director of STEM
Massachusetts Department of Elementary and
Secondary Education
Former Senior Program Director for Science
Boston Public Schools
mjrdecker@gmail.com

Chapter 11

Elaine Woo
Elaine Woo Consulting
3001 120th Avenue NE
Bellevue, WA 98005
elainewoo@aol.com
425-881-1810
(Formerly the P–12 Science Program Manager
in Seattle Public Schools)

Chapter 12

Nancy (Anne) Degnan, PhD
Executive Director
Director, Academic Initiatives
Office of Academic and Research Projects
Center for Environmental Research
and Conservation
Earth Institute, Columbia University
475 Riverside Drive
New York, New York 10027
ald1@columbia.edu
212-854-8310

Chapter 13

Jane B. Huffman, EdD
Professor
Teacher Education and Administration
University of North Texas
jane.huffman@unt.edu
940-565-2832

INDEX

*Page numbers printed in **boldface** type refer to tables or figures.*

INDEX

INDEX

INDEX

R

INDEX